AN ABUNDANCE *of* FLOWERS

AN ABUNDANCE *of* FLOWERS

more great flower breeders
of the past

Judith M. Taylor

SWALLOW PRESS / OHIO UNIVERSITY PRESS
Athens, Ohio

Swallow Press
An imprint of Ohio University Press, Athens, Ohio 45701
ohioswallow.com

© 2018 by Judith M. Taylor
All rights reserved

To obtain permission to quote, reprint, or otherwise reproduce or distribute material from Swallow Press / Ohio University Press publications, please contact our rights and permissions department at (740) 593-1154 or (740) 593-4536 (fax).

Printed in the United States of America
Swallow Press / Ohio University Press books are printed on acid-free paper. ♾ ™

28 27 26 25 24 23 22 21 20 19 18   5 4 3 2 1

Cover photograph: Clematis 'Nikita'.
Photograph by Sergei Afanasev, Shutterstock.com
Cover design: Chiquita Babb

Library of Congress Cataloging-in-Publication Data

Names: Taylor, Judith M., author.
Title: An abundance of flowers : more great flower breeders of the past / Judith M. Taylor.
Description: Athens, Ohio : Swallow Press/Ohio University Press, [2017] | Includes bibliographical references and index.
Identifiers: LCCN 2017043956| ISBN 9780804011921 (hc : alk. paper) | ISBN 9780804011938 (pb : alk. paper) | ISBN 9780804040853 (pdf)
Subjects: LCSH: Flowers--Breeding--Europe--History. | Flowers--Breeding--United States--History. | Plant breeders--Europe--History. | Plant breeders--United States--History.
Classification: LCC SB406.8 .T37 2017 | DDC 635.9--dc23
LC record available at https://lccn.loc.gov/2017043956

FOR BARBARA AND PHIL

In affectionate friendship

CONTENTS

INTRODUCTION	1
CHAPTER 1. Poinsettia	5
CHAPTER 2. Chrysanthemum	19
CHAPTER 3. Penstemon	65
CHAPTER 4. Gladiolus	81
CHAPTER 5. Dianthus (Carnations and Pinks)	113
CHAPTER 6. Clematis	143
CHAPTER 7. Pansy/Viola	167
CHAPTER 8. Water Lily	189
AUTHOR'S NOTE	207
ACKNOWLEDGMENTS	209
NOTES	213
REFERENCES	215
INDEX	223

AN ABUNDANCE *of* FLOWERS

Introduction

ALL OF US, gardeners and nongardeners alike, are extraordinarily spoiled in being able to satisfy almost any floral whim with a huge abundance of choices. Did you ever wonder about the amazing variety of potted and cut flowers found even in the local supermarket, why you have roses which start to bloom in the early spring and continue late into the year, or why there are chrysanthemums in almost every color of the rainbow, including green? How about that prosaic garden center or "big box store" down the street and its vast inventory? Did you ever ask yourself how they could offer such varied flowers and who might have created them?

It is hard to imagine a time when florists' carnations were new and exotic, but that was the case toward the end of the nineteenth century and into the early twentieth century, when the humble pink became a great star. Begonias are another example. The new flowers came in waves and created great excitement. Many began as rich men's playthings, like orchids, but rapidly came down in price and spread to the general population.

There is hardly a flower or shrub in general use that has not been crossbred and hybridized since its original discovery. In *Visions of Loveliness: Great Flower Breeders of the Past,* I touched on the fascinating stories behind eighteen well-known flowering plants, but no work of that sort is ever complete: One simply draws a mental line at a particular point and stops. As soon as the manuscript has been sent to the publisher, new material appears, and the task stretches out asymptotically.

With that in mind, I decided to pick up where I had stopped and cover another series of handsome plants with equally fascinating stories. In this book, I look into several more herbaceous flowers and one vine.

In a few instances, figures familiar from the first book make another appearance. Victor Lemoine (1823–1911) is one of them. Lemoine was the son and grandson of well-established estate gardeners in Alsace-Lorraine. The family worked for a wealthy nobleman. It was a sign of their prosperity and social position that Victor Lemoine's parents could afford to send him to a good school and allow him to stay there until he was seventeen. Usually, gardeners' children were lucky to go to school at all. It was clear Victor Lemoine knew exactly what he wanted to do, because he apprenticed himself to three of the most influential horticulturists of his day. Within a year of opening his own nursery in 1844 in Nancy, a rapidly growing industrial city in eastern France, he issued the first of his intentional hybrids, a new purslane. Shortly after that, more of his hybrids drew attention and he was the subject of an article in a French horticultural journal.

During his self-directed apprenticeship, Lemoine spent about a year in Ghent with Louis Van Houtte, a now-legendary figure who expanded the radical new business of breeding flowers. Van Houtte had a very powerful vision of what could be done and evidently found backers to support it. Ultimately Van Houtte covered acres of land in Ghent with greenhouses. During that epoch, Belgium led Europe in the flower business, equal at least to the industry in the Netherlands. Lemoine looked and learned.

In this volume, we encounter Lemoine introducing a dizzying number of penstemons, almost 500 cultivars. He did not do so badly with gladioli either, nor with pelargoniums, begonias, peonies, deutzias, and weigelas, yet even he was overtaken by another one of those industrious Scots whose energy and drive are legendary. In the modest provincial town of Hawick, Scotland, John Forbes developed 550 cultivars of penstemon. A few of his introductions are still available. It turns out that Forbes corresponded with Lemoine and used plants from Nancy as the basis of some of his crosses.

With more information resulting from persistent research, a pattern is recognizable. For every group of modest horticulturists who did fine work and left a small legacy, there were indeed formidable figures who once were feted and admired but whose names have languished unknown in recent years. It may seem pettifogging and unnecessary to bring back so many minor breeders, but the reward lies in restoring the reputations of almost larger-than-life horticulturists.

In nearly every chapter, the reader will find a heartening story of a flower breeder who built a vast business, supplying the rest of the United States with seeds and bulbs and even sending his wares abroad. As long as they lived, the businesses flourished, but unless they had children who took over, or devoted partners who continued after they died, everything evaporated. If ever the tag "*Sic transit gloria mundi*" applied, it is to the flower breeding business. These people are merely dusty echoes now.

Luther Burbank's reputation is secure because of the scale of his achievements and the importance of his work. Burbank bred the formidable potato that changed the way this crop was grown and was a huge benefit to humanity. He also introduced many new types of fruit as well as some flowers such as the Shasta daisy, still a valuable ornamental plant. Because of these accomplishments, Burbank has become a legend.

Few can compete with that. Great flower breeders like Henry Harris Groff and his gladioli and Wilhelm Pfitzer with gladioli, penstemons, and several other flowers, for example, had triumphant achievements in their lifetimes. Pfitzer and his descendants won so many medals and cups that a numismatist wrote an entire book about them. But no one remembers Pfitzer or Groff today apart from relatives.

Such accomplishments have been absorbed into the sum total of horticultural knowledge as the world has moved on. The individual breeders have been forgotten. Many were extremely prosperous and well-known in their time, but now they have to be sought in archives. It is intensely gratifying to be able to restore their lives and reputations.

1

Poinsettia

WHEN THIS project began, I asked a distinguished horticultural editor if he would like an article on the development of the poinsettia. Until then he had put up with my peculiarities and indulged me, but here he drew the line. "Poinsettias," he said firmly, "are not plants any longer, they are a commodity." He truly believed what he said, but oh, how unfair.

The beautiful poinsettia, known for its scarlet bracts, comes to us encrusted with myth and legend, as befits a royal plant of the Aztecs. The Nahua people in Mexico called it *cuetlaxochitl* (xochitl is the ancient Nahuatl word for an ornamental flower). Poinsettia is a desert plant and very sensitive to the cold. As Tenochtitlan (now Mexico City) was at high altitude, poinsettia did not flourish there, but every winter the rulers imported thousands of the plants from warmer regions. Extracts of the plant were used to dye cloth, and its latex was used for medicinal purposes. The Spanish conquerors and missionaries attempted to erase all evidence of the preceding pagan Aztec religion, but records have survived showing that the plants were used for religious ceremonies in the winter.

Once the Spanish friars took over, they adopted the brilliant red plant as part of the Christmas ritual, and the Spanish-speaking Mexicans named it *flore de nochebuena*, the flower of the Holy Night (Christmas Eve). The vivid red bracts of poinsettia, which emerge in early winter, have signified the festive mood at Christmas and the joy of the season for over 150 years in the United States and Europe. In what follows, I hope to establish the actual story of its arrival in the United States and Europe and its extraordinary development, and will attempt to clear away all the accumulated misinformation and cobwebs.

BOTANY

The poinsettia (*Euphorbia pulcherrima* Willd. ex Klotzsch) is a member of the large and diverse family Euphorbiaceae. The plant originated in southern Mexico and northern Guatemala. In its native habitat, this species is a winter-flowering shrub that grows over three meters high and is a common landscape plant. The sap is milky and may produce dermatitis in susceptible individuals. The umbel-like cymes are subtended by many showy bracts, usually red, but breeders have used their imagination to produce many different colors, including white, pink, and purple. In part this has been driven by the demands of the market.

Other major innovations in the poinsettia resulted from the remarkable discovery by Gregor Gutbier, an Austrian poinsettia breeder in the 1980s, that grafting poorly branched plants onto well-branched plants increased branching in the propagules of the restricted-branching plant. This effect was demonstrated to be due to the transmission of a phytoplasma from an infected to a healthy plant. The phytoplasma was later shown to be similar to the infectious agent causing peach X-disease and spirea stunt but acted in a benign manner in poinsettia. Once the role of the virus was recognized, it became standard procedure to introduce the beneficial pathogen to new poinsettia seedlings by grafting. Propagules from grafted plants kept their free-branching trait. Other innovations in poinsettia production include pinching to increase branching, and the use of growth regulators to reduce plant size.

Species euphorbias in their native habitat. Artist unknown.
Reproduced by permission of Chronica Horticulturae

Joel Roberts Poinsett.
Artist: Charles Fenderich (1838).
Reproduced by permission of National Portrait Gallery, Smithsonian Institution

JOEL ROBERTS POINSETT AND THE POINSETTIA PLANT

The common name, poinsettia, honors an American, Joel Roberts Poinsett (1779–1851), who saw the plant in southern Mexico in 1828. For years it has been assumed that Poinsett came across the gorgeous plant in Taxco in southern Mexico, and took it home to Charleston, South Carolina, in 1828. That is not correct.

From there he is said to have sent cuttings to Colonel Robert Carr, a nurseryman in Philadelphia, Pennsylvania, whose wife, Ann, was the granddaughter of John Bartram, the self-taught American botanist of the colonial era. In June 1829, Carr entered the plant as "a new *Euphorbia* with bright scarlet bracteas or floral leaves, presented to the Bartram Collection by Mr. Poinsett, United States Minister of Mexico," at the Pennsylvania Horticultural Society's flower show, where it was seen and admired by hundreds of people.

Robert Buist, a Scottish nurseryman in Philadelphia, took the next step in its dissemination. He was so enthralled by the new plant that he took cuttings to his friend James McNabb in Edinburgh. From Scotland, it reached the distinguished German botanist Karl Willdenow in Berlin, who named it *Euphorbia pulcherrima*, in 1834. This remains the accepted botanical name. Two years later, Robert Graham in Edinburgh published his taxonomic findings and changed the name to *Poinsettia pulcherrima*.

Another myth pertains to how Willdenow came to study the plant. It is said that the plant somehow crept into his greenhouse through a hole in the wall, but the present director of the Berlin-Dahlem Botanical Garden, Dr. H. Walter Lack, told me that he considers this story to be completely false.

Charming and delightful as this tale of the poinsettia might be, the actual facts are as follows: The plant is endemic to southern Mexico. Specimens arrived in the United States in 1828, and by 1829 it was on display in Philadelphia. Its arrival was associated with the name of Joel Roberts Poinsett. From Philadelphia, it crossed to Scotland, and then to Germany. There is no evidence that the plant was ever in Charleston, South Carolina, before reaching Philadelphia, but its movements after Philadelphia are well documented.

When J. Fred Rippy wrote a biography of Poinsett in 1935, he devoted one paragraph to the poinsettia story and indicated in a footnote that he had been unable to find any correspondence to validate the claim that Poinsett had introduced the plant. He commented drily that "it is generally acknowledged in the horticultural guides that Mr. Poinsett introduced the flower."

Rippy cited the only reliable document of the era, a discussion of Poinsett in the *1887 Charleston Yearbook* by Charles Stille, who had spent a day with Poinsett as a lad of twelve. Poinsett took the young boy with him when he visited the Reverend John Bachman, a Lutheran minister and noted naturalist who once worked with Audubon. The *Yearbook* article states: "Mr. Poinsett was rewarded for the interest he took in science by having a beautiful flower named after him." There is some "difference of opinion as to whether Mr. Poinsett discovered it himself or simply introduced it to this country. At all events it is always known now as being named after him."

At the time, the flower was called either "Mexican flame flower" or "painted leaf" in the United States. Neither of these seemed satisfactory, and this is a rare occurrence of a plant acquiring an enduring common name after it received its formal name, rather than the other way around. The choice of Poinsett's name is often attributed to William Hickling Prescott, the author of the classic book *The History of the Conquest of Mexico,* but this too is a myth, since Robert Graham used the name *Poinsettia* in his taxonomic identification of 1836.

Poinsett was a very well-educated, cosmopolitan Southern gentleman of Huguenot descent from Charleston, South Carolina, who spoke French, German, Italian, and Spanish. He was appointed the first American minister to the newly independent Mexico by President James Monroe in 1825, but was recalled by President Andrew Jackson in 1830. Poinsett subsequently acted as secretary of war in President Martin van Buren's cabinet after serving terms in the South Carolina state legislature and the U.S. House of Representatives. He wrote a book about his first tour of duty in Mexico, *Notes on Mexico* (1822), with no mention of the plant.

Poinsett never enjoyed very robust health. He started out to be a physician like his father but could not complete the course. His lifelong interest in natural science stemmed from the preliminary studies he did finish. While in Mexico, Poinsett carried on an extensive correspondence about horticulture, exchanging seeds and cuttings with friends and colleagues in the United States. He also believed that the exchange of plants and seeds helped to promote stronger ties between the United States and Mexico.

Joel Fry notes that the American Philosophical Society in Philadelphia elected Poinsett to membership in 1827. This broadened his correspondence to include members of the society and other Philadelphia savants. These connections appear to be the most likely route through which the new plant with bright red bracts reached the United States. There are fairly strong indications that it traveled directly from Mexico to Philadelphia, as four different collections of Mexican seeds and plants were dispatched to Philadelphia between 1828 and 1829.

William Maclure, the president of the Academy of Natural Sciences of Philadelphia, and Thomas Say, a descendant of John Bartram, visited Poinsett in Mexico for three months in January 1828, traveling to Veracruz and Mexico City. Later that year, Maclure visited Poinsett again, and returned to Philadelphia in the fall with many seeds and plants. Say also collected more than a hundred types of seeds but was not meticulous about identifying them. Number 65, a "Fine Red flower, perennial," could be poinsettia.

In November 1828, James Ronaldson, a Scottish enthusiast in Philadelphia, wrote to Poinsett, who had remained in Mexico until 1829, that he had received a box of seeds from Veracruz and assumed it had come from him. The fourth possibility was William Keating, a geologist who went to prospect in Mexico and met Poinsett. On occasion, Keating acted as Poinsett's courier.

In summary, there is no doubt that the plant was being grown in Philadelphia when Colonel Robert Carr exhibited it at the first flower show of the Pennsylvania Horticultural Society in June 1829. Poinsett was still in Mexico, but it was generally accepted that he had had a lot to do with the plant reaching the United States. Perhaps the following extract from a letter from one of Poinsett's friends in 1830 clinches the argument that these plants did not enter the United States via South Carolina. The letter discusses a woman from Charleston: "Mrs. Herbemont has *been very vexed* with you when she learned by the papers that several northern gardeners had received seeds and plants you had sent them from that land of vegetable beauties, Mexico, and that *you had not in one instance* remembered her."

The Poinsettia in Mexico

The specimen received in Philadelphia was not a wild plant but had been cultivated and modified for many years in its native Mexico. Doña Fanny Calderon de la Barca, wife of the Spanish minister to Mexico, commented in her letters home that her church courtyard was lit by these gorgeous scarlet flowers at Christmastide. For reasons that are not clear, Mexican growers still believe that Poinsett devised a hostile mechanism to prevent them from developing or benefiting from the plant's popularity, purely out of spite. Various publications in Mexico state that Poinsett obtained a "patent" in the United States, which led to this embargo.

Numerous scholars have searched through old patents and treaties but failed to turn up such an instrument. Although U.S. patent laws began in 1795 to protect inventors against their mechanical devices being pirated, these laws did not cover plants. The first U.S. law that did protect new cultivars of plants, the Townsend-Purnell Act of 1930, excluded seed-propagated plants, tuber-propagated plants (other than potato), and wild plants.

At present, international protection for plants is controlled by treaty (International Union for the Protection of New Varieties of Plants, or UPOV, adopted in 1961). Seed-propagated plants in the United States are protected by the Plant Variety Protection Act of 1970, administered by the U.S. Department of Agriculture (USDA). Poinsett did negotiate a commercial treaty with Mexico as part of his ministerial duties, and it was ratified by the United States, but the poinsettia was not part of the treaty. Modern twentieth- and twenty-first-century poinsettia cultivars are protected by patents, but these are relatively recent advances.

Mexican animosity toward Poinsett has some basis in fact and this may have contributed to the myth of the U.S. patent. Poinsett was a very upright, even self-righteous, man and took his duties seriously. This led him to meddle in Mexico's internal affairs, supporting one party over another, clearly an infraction of diplomatic rules. At one point, there were even death threats against him. All this contributed to his recall by the American president. The term *poinsettismo* is still in use today in Mexico to express arrogance and high-handedness.

The Poinsettia in the United States

North American nurserymen began to propagate the poinsettia rapidly and distributed it widely throughout the United States over the last part of the nineteenth century. The modern phase of poinsettia development took place in the United States by the early twentieth century. Poinsettias have led the sales of potted plants year after year, and the poinsettia is now one of the mainstays of the commercial flower market. This phenomenal growth is associated with the Eckes, a German immigrant family that settled in southern California in 1900.

Albert Ecke and his family stopped over in California in 1900, en route to Fiji, where they planned to open a health spa. They saw such an excellent opportunity in California that they settled there instead, with their descendants remaining through the present day. Albert began farming in the Eagle Rock Valley, near Los Angeles, but then moved to Hollywood. The family planted orchards and also large fields of chrysanthemum, gladiolus, and poinsettia for the cut flower market. By 1909, they had narrowed their floral crops down to poinsettia alone. Ten years later, both Albert and his eldest son, Hans, had died, and the business was taken over by the second son.

Paul Ecke Sr. moved it south to Encinitas, where it remains today. Paul Ecke Sr. found some valuable spontaneous hybrids in his open fields, but it was his son, Paul Ecke Jr., who initiated a scientific breeding program. At one

Three generations of the Ecke family: Paul Ecke (son of Alfred Ecke), Paul Ecke Jr., and Paul Ecke III.
Reproduced by permission of Paul Ecke III

point the business was sold to a Dutch horticulture firm, but Paul Ecke III has bought it back and is restarting it.

The history of the poinsettia in the United States in the twentieth century has some well-defined landmarks. Major advances came about with the discovery of photoperiodism in plants by Wightman Garner and Henry Allard (1920). This led to the use of opaque black cloth to shorten day length. This was essential because the poinsettia is a short-day plant. The plant flowers when darkness lasts for at least 11.75 hours. It is the length of the night, not the day length, that is critical. Adding lights to interrupt the dark period prevents flowering. Shortening the day length and increasing the dark period induces flowering in the poinsettia. Increasing the day length by supplemental lighting keeps the plant vegetative. Management of day length permits synchronization of flowering in order to get plants to flower for the Christmas season. In all this one must remember that the colorful bracts are the major feature of the plant, not the tiny yellow blossoms in the center. The term "flowering" is used loosely.

The early poinsettias were still very fragile. Their leaves fell off quickly, and the scarlet bracts only lasted for about a week to ten days. It made it very difficult to get them into perfect condition by Christmas. That remained true until the 1950s. The late Lyndon Drewlow, a well-known breeder of poinsettias, recalled that he had to assist his professor at graduate school with shortening the stems manually and performing other maneuvers to get the plants into condition for seasonal sale (Drewlow, personal communication).

Radically different cultivars of poinsettias became available over time and changed the direction of its development. Some of this was due to the establishment of a number of breeding programs across the country in the mid-1950s, including Pennsylvania State University; the USDA Research Center at Beltsville, Maryland; and the University of Maryland. Private companies like Azalealand in Lincoln, Nebraska; Mikkelsens in Ashtabula, Ohio; Earl J. Small of Pinellas Park, Florida; and the Yoder Brothers in Barberton, Ohio, were also very active.

One USDA geneticist, Dr. Robert N. Stewart, separated out the most desirable characteristics, such as large bracts, stiffer stems, new colors, and the ability to last for a longer time, and bred for these. The key cultivars were 'Oak Leaf' (1923); 'Paul Mikkelsen' (1963); 'Eckespoint ® Lilo' (1988); 'Eckespoint Freedom' (1992); and 'Eckespoint Winter Rose Dark Red' (1998).

At first the Eckes only grew the two cultivars of poinsettia available before 1920: 'True Red' and 'Early Red'. Their neighbors in Southern California used these plants to ornament their gardens. 'Early Red' was more useful for commercial purposes both as a cut flower and as a potted plant because it held its foliage longer. Three new cultivars were released in the 1920s, but just one of them, 'Oak Leaf', introduced by a Mrs. Enteman in Jersey City, New Jersey, dominated the field for the next forty years.

It was the first cultivar suitable for growing in a pot and also retained its leaves and bracts for a longer time. The other two, a 1920 sport, 'Hollywood', with wider, more compact bracts than 'Early Red', and the 1924 'St. Louis' from Louis Bourdet in St. Louis, Missouri, did attain some popularity in their day. Paul Ecke devoted himself to selecting and developing better cultivars based on 'Oak Leaf'. His introductions included 'Henriette Ecke' (1927) and 'Mrs. Paul Ecke' (1929). The latter, a sport of 'Oak Leaf', was shorter and had wider bracts than its parent.

Peach poinsettia.

By the late 1920s, poinsettias had become a commercial reality, and several firms across the United States grew them successfully in greenhouses. In Indianapolis, Baur and Steinkamp came across another sport of 'Mrs. Paul Ecke', which they named 'Indianapolis Red'. Each of these sports offered improvement in habit and bract size.

Not all the new cultivars lasted well, despite their undoubted novelty. For example, the offspring of 'Henriette Ecke', a cultivar with "double" incurved bracts, which made the plant almost look like a dahlia, seemed very promising, but the bracts were deemed to be too small and the plants did not perform well in the greenhouse.

Demand has since risen for novelties, such as 'Winter Rose Dark Red' (1998), the first cultivar to have "curly" incurved bracts and very dark, incurved foliage. By 2004 it was available in seven different colors. Another series with curly bracts, Renaissance, came in at about the same time, specifically for the cut flower market. These varieties do very well as cut flowers.

Although the public finds new colors and styles very exciting, the traditional red poinsettia has remained popular, with desirable traits continuing to be developed. 'Paul Mikkelsen' from the Mikkelsen nursery in Ashtabula, Ohio, had a stiffer stem and greater longevity than any preceding cultivar. Eckespoint 'Lilo' was the first poinsettia with dark leaves and early flowering. It retained its foliage well but needed some special treatment to ensure good branching. Eckespoint 'Freedom' had all the above good points but more consistent branching. It was also ready to be shipped a week or two before Thanksgiving, allowing for a head start on the holiday marketing season. Another excellent quality was its ability to withstand careless handling by untrained staff at large nonspecialty stores.

Breeders have to respond to these needs and accommodate the public's slightly fickle reactions. Since 2002, Ecke has introduced Eckespoint 'Plum Pudding', with purple bracts; Eckespoint 'Chianti', with darker wine-red bracts; Eckespoint 'Shimmer Pink', pink with white flecks; and many others. Eckespoint 'Prestige Red' has become the standard modern cultivar, and it already has many variations.

Poinsettias are no longer grown in the continental United States on a large scale. Almost all cultivation is now done in Central and South America. It is cheaper to grow the plants there because the climate is warmer and the cost of labor is lower.

The Poinsettia in Europe

The poinsettia was widely distributed across Europe by the mid-nineteenth century. It enjoyed great popularity for the same reasons it was so successful in North America, but because the plant had to be grown in heated greenhouses, it was an expensive luxury. Thormod Hegg, a Norwegian breeder, introduced 'Annette Hegga Red' in 1964. Hegg found that pinching the stem during development stimulated the growth of more than one inflorescence. There could be as many as eight of them. Hegg's cultivars also came in a previously unknown range of colors. In addition, it was easy to propagate and grow commercially. The line lasted until 2002.

The Zeiger Brothers in Hamburg, Germany, instituted a breeding program. Gregor Gutbier, in Linz, Austria, introduced another dazzling series of colorful cultivars, the V-14 Glory Angelikas, in 1979. Ten years later, these plants reached the United States. One advantage they offered was an ability to withstand slightly cooler night temperatures. Gutbier was the first grower to realize that grafting an attractive plant with good bracts and branching onto a less well branched plant led to more uniform and successful results. Grafting was very important during the 1980s. Once the role of the virus was recognized, many more breeders were able to produce reliable new cultivars, and the market shares were redistributed.

2

Chrysanthemum

CHRYSANTHEMUM HISTORY, like Gaul, is divided into three parts. The first part is the epoch of extensive and intensive cultivation in China, and later in Japan, lasting about 1,400 years from the fifth century BCE. The author of a recently published book about the origin of the chrysanthemum, J. J. Spaargaren, an eminent Dutch horticulturist, has elucidated this period thoroughly for Western scholars.

The second part was the initial introduction of the flower into Western Europe in the late seventeenth century, when very little happened. No one paid much attention to the modest flower and it lapsed into obscurity. The third part, the one covered in this chapter, was the beginning of the modern period in 1789, when Captain Pierre-Louis Blancard (not Blanchard as is sometimes written) of Marseilles imported specimens of *Chrysanthemum morifolium* from China and gave cuttings to the Abbé de Ramatuelle. This period continues into the present.

Chrysanthemum morifolium was the first really large-flowered chrysanthemum ever seen in Europe. Only a few had survived the journey, and the

Chrysanthemum morifolium 'Old Purple'.
Reproduced by permission of Chronica Horticulturae

one Blancard sent the abbé, which ultimately reached London, was tall and purple. The first European illustration of this new flower was plate 327 in the *Botanical Magazine* in 1796. The abbé, in his turn, sent specimens to the Jardin du Roi in Paris in 1790.

The survival of this fragile new flower during those perilous times in France is astonishing. With the revolution raging, it is remarkable that people even thought about flowers. It is even more amazing that the Jardin du Roi, a symbol of the hated royal family, escaped utter destruction. The institution, which actually flourished under the new regime, was renamed the Museum of Natural History in 1800. A few of the more intellectual of the revolutionaries believed that understanding natural history was important to a new world order.

BOTANY

Chrysanthemums are in the family Asteraceae. This family produces blossoms with large composite heads. Short disc florets are clustered in the center of the blossom, and long, slim ray florets supply the "petals" surrounding the disc. The current botanical name of the horticultural chrysanthemum is *Dendranthema grandiflorum,* but in the United States it goes by the affectionate nickname of "mum." When the name *Dendranthema* was proposed to replace *Chrysanthemum,* the members of the English National Chrysanthemum Society complained so vociferously that the change did not take place. Germplasm from other *Dendranthema* species, possibly *D. japonicum* and *D. indicum,* also contributes to the modern hybrid. *C. indicum* is the type species of the genus *Chrysanthemum.*

The next important division is into classes based on the flower's shape, the so-called horticultural divisions. This nomenclature has evolved over the past century, absorbing new forms as they have been discovered. The classification system of the Chrysanthemum Society of America is based on the shape and arrangement of the rays as well as the disc florets: Irregular Incurve; Reflex; Regular Incurve; Decorative; Intermediate Incurve; Pompon; Single/Semi-Double; Anemone; Spoon; Quill; Spider; Brush and Thistle; and Exotic. These classes are in keeping with the international system.

The pompon variety was originally quite tall and even needed staking in some instances. The flower's resemblance to the pompon on top of the French sailor's hat was irresistible, and the name stuck. For a while, the early growers called it "pompone." As members of the Asteraceae, chrysanthemums and dahlias look very much alike, up to and including the pompon forms.

The plant is native to the Northern Hemisphere and widespread across the Eurasian landmass. It is found most abundantly in the Mediterranean region, particularly Algeria and the Canary Islands, and in northern Asia: China, Japan, and Korea. A few species, mainly in the genus *Tanacetum*, are endemic to North America. The European and North African species are diploid, whereas the Asian plants range from diploid to decaploid. ("Diploid" means there are two sets of chromosomes. "Decaploid" means there are ten pairs. The presence of more chromosomes is often associated with larger and more complex flower heads.)

COMMERCIAL IMPORTANCE

This flower is now one of the most important floricultural crops in many countries. Together with poinsettias and orchids, the chrysanthemum appears at various positions in the lists of the top ten most frequently sold potted plants, cut flowers, and garden plants in the United States and other countries. In the 1990s, Japan led the way with 2 billion stems of cut flowers. In contrast, during that period, the Netherlands sold 800 million stems, Colombia 600 million, Italy 500 million, and the United States 300 million. Chrysanthemums in Italy are almost solely used for funereal purposes, and Italians get quite upset if a guest arrives with a bunch of chrysanthemums as a hostess gift.

In the U.S. Department of Agriculture data for 2009/2010, the grand total of hardy potted chrysanthemums sold was 45 million pots, with 7 million indoor pots, as well as 8 million bunches of cut pompons. (By comparison, 36 million pots of poinsettias were sold during that same period, and 21 million pots of orchids.)

DEVELOPMENT IN EUROPE

Chrysanthemums seized imaginations in England and France at much the same time. Here was a compact floriferous plant, available in attractive colors, easy to grow, and coming into bloom at the end of the summer. It could continue in flower as late as December. This was something to conjure away the bleak dullness of late autumn. An English horticultural observer, A. H. Haworth, suggested that if they were planted against a sunny wall in England and properly tended, the flowers might still be blooming in January. The French climate was somewhat more propitious, and the flowers flourished in the warmer regions.

ENGLAND

The chrysanthemum was said to have flowered for the first time in England at Colvill's nursery in the King's Road, Chelsea, in 1796. Colvill's 'Old Purple', a form of the purple flower Blancard had originally given to the Jardin du Roi in France, was tall with double purple flowers. Its Linnaean name was *Chrysanthemum morifolium* (now *C.* × *morifolium* Ramat). Chrysanthemums had been grown at the Chelsea Physic Garden many years before, in the quiescent "second epoch," but they had been ignored and lost.

A Dutch merchant, Jacob Layn, had introduced chrysanthemums into the Netherlands in about 1688, but just as occurred in England, once they died out no one seemed to remember anything about them, and a century later it was as if they had never been there. There were said to be six varieties of the flower. William, Prince of the House of Orange in the Netherlands, took over the throne of England in 1688 as William III. He and his wife, Mary, introduced the Dutch style of gardening. They reigned together, always referred to as "William and Mary," and favored trees and shrubs over herbaceous plants. The chrysanthemum was not used.

After the flowers were reintroduced, the chrysanthemum began to spread slowly throughout England. The Horticultural Society of London was enthusiastic, and *Curtis's Botanical Magazine* had quite a few pictures of this "new" flower. George Harrison of Downham in Norfolk was especially enamored of it. In 1831, he protected his late plantings under glass, with stunning results. Eventually his efforts led to the first chrysanthemum show, held in Norwich in 1843.

Three years later, the Stoke Newington Chrysanthemum Society was started. This later became the National Chrysanthemum Society. Stoke Newington was a charming village just outside north London at the time. It has since been incorporated into north London.

In his 1971 book about antique plants, Roy Genders mentioned several chrysanthemum enthusiasts who were active before the next major event, the advent of the Chusan Daisy variety. These included Isaac Wheeler of Oxford, who exhibited his flowers at the Horticultural Society in 1832, and another resident of Downham, John Freestone of Watlington Hall, who was the first Englishman to ripen seed and raise new varieties. Chrysanthemum seed is hard to collect and handle, so this was a real achievement.

Vauxhall, in London, the home of the famous and also infamous Vauxhall Gardens (who knows what high jinks went on there), was where Chandler's Nursery grew seedlings from seed sent by John Salter. Salter was an English nurseryman who worked in Versailles for a number of years before having to return to England in 1848. The populist uprisings in Paris made things too uncertain for him. Two of Salter's cultivars, 'Queen of England' and 'Annie Salter', lasted a very long time and could still be found in some nursery catalogues as late as 1960.

Early enthusiasts also had *C. indicum* in their gardens. Its single yellow flowers provided an additional source of color. Its origins went back almost a century. In 1751, Peter Osbeck, one of Carolus Linnaeus's students, had found *C. indicum* near Macao in southern China and sent it back to Europe. Philip Miller cultivated it in the Chelsea Physic Garden as early as 1764.

In 1822, J. C. Loudon, the formidable one-armed horticultural editor and writer, said that Joseph Sabine, secretary of the Horticultural Society of London (subsequently the Royal Horticultural Society), knew of fourteen types of chrysanthemum. Loudon also commented that there were supposed to be more than fifty types of chrysanthemum in China. By 1826, Sabine could point to forty-eight varieties of this plant in the society's grounds. In that year Louis Noisette, noted for his roses, took a few of Sabine's varieties back to France. Years later, Sabine insisted on sending Robert Fortune to China to bring back even more chrysanthemums.

Sabine's list of chrysanthemums at the Horticultural Society of London in 1822 included these varieties:

Buff Flowered	Changeable White
Golden Yellow	Large Lilac
Pink Flowered	Quilled Pink
Quilled Flamed Yellow	Quilled Yellow
Quilled White	Spanish Brown
Sulphur Yellow	Superb White
Tasselled White	The Purple

Until Robert Fortune brought the Chusan Daisy, *C. rubellum*, back from China in 1846, these *C. indicum* and *C. morifolium* varieties were the only types of chrysanthemum in the British Isles. They formed the backbone of all breeding efforts. You will look in vain for *C. rubellum*. It is now *Chrysanthemum zawadskii* subsp. *latilobum* (Maxim.) Kitag.

Fortune's introduction became very popular, and led to considerably increased interest in the flower. The diminutive daisy-like plant was the forerunner of many new cultivars. Another great advantage of the Chusan Daisies was that they flowered much earlier in the year, enabling them to be grown outdoors.

It was still possible to find some of the British varieties from the early and mid-nineteenth century in the twentieth century. Genders listed at least three pompons which appeared in John Forbes's catalogue in 1960: 'Model of Perfection', 'Bob', and 'Mlle. Marthe'. Forbes had a nursery in Hawick, Scotland. (See chapter in this book on the penstemon.) Robert Fortune embellished his already stellar reputation for finding magnificent plants by collecting Japanese varieties and taking them back to England in 1862. They were quite unlike the previous specimens. Some were shaped like a camellia, and there was a wider range of colors. According to J. Lochot, himself a chrysanthemum breeder, Fortune's introduction of seven Japanese species in 1862 invigorated the field; everyone was excited by the large blossoms with long, narrow, and fantastical petals.

Putting all these things together, nurserymen were able to establish a successful commercial cut flower trade in chrysanthemums.

Wealthy men of leisure as well as nurserymen devoted their lives to growing and breeding chrysanthemums. They formed societies of like-minded people, held competitions, and moved the flower in exciting new directions. The societies created increasingly complicated and strict rules to govern the exhibitions, constantly tightening the challenge. Rigid rules are a feature of English floral competitions and shows, sometimes stultifying genuine advances and necessary change. The Royal Horticultural Society (RHS) was particularly important in promoting these improvements. Members who disagreed and amateurs were the ones who restricted new advances. This conservative attitude was a feature of the old workingmen "florists," mentioned below.

The RHS established a Floral Committee in 1859 to report on all flowers or flowering plants being submitted for consideration. Looking at the members of this committee, one is struck by the number of active nurserymen, some of them chrysanthemum breeders, who served on the committee. Although the society was an elite social organization, the fact that its members decided to include these tradesmen was a sign of how seriously they took horticulture. The leaders were able to transcend their class consciousness and recognize true talent and contributions when they saw them. As a complete *non sequitur*, it is almost certain their wives did not approve. Well-born women were constantly aware of class distinctions.

Florists and the Chrysanthemum

For almost three hundred years, working-class men in the British Isles found sources of pride and dignity in growing and breeding a number of small plants like auriculas, pinks, carnations, primulas, tulips, and ranunculus, which lent themselves to containers. This activity started when the Huguenots (French Protestants) fled from Paris after the infamous Saint Bartholomew's Day massacre in 1572. The refugees took their cherished plants with them, and the idea of plant breeding spread. These men were very poor, without estates or property, but they still enjoyed the excitement of watching plants grow and selecting new varieties. They also enjoyed competing for prizes, and, perhaps even more, the respect that winning brought them. After some delay, the chrysanthemum was also taken up by these "florists," as they were called.

Florists tended to be small artisans such as skilled cotton weavers in Lancashire and the silk weavers of Paisley. They were able to tend their looms and keep an eye on their flowers at the same time. If it began to rain, they could take the cherished auricula with its delicate "farina" inside the house.

Once the crucially important textile industry was consolidated into large factories in the early nineteenth century, this convenient state of affairs was no longer feasible. Proudly independent weavers became cogs in a gigantic machine. A factory hand can only take care of plants in his free time. Florists and their societies gradually disappeared over the nineteenth century.

Florists adopted the 'Gold Bordered Red' variety of chrysanthemum in about 1830. The color, gold-tipped red petals and an undersurface striped with gold, and form both suited their slightly odd attitudes toward beauty. Until then they had been indifferent to the flower's charms. A flower all in one color was not challenging. The florist wanted two or more colors to stimulate his fancy. As A. H. Haworth wrote in J. C. Loudon's *Gardener's Magazine* in April 1833, "Chinese chrysanthemums have not hitherto ranked with the true flowers of the florists because, however well-formed, in many of the varieties, they are all, save the Gold-bordered Red, of self or uniform colours." A flower with gold-tipped red petals and an undersurface striped with gold did catch their eye. This feature allowed the grower to come up with stripes, flakes, and picotees and generally follow his whims.

Classification

Mr. Haworth also tried his hand at a classification in the same article. He listed seven classes: Ranunculus-flowered, Incurved Ranunculus-flowered, China Aster-flowered, Marigold-flowered, Tassel-flowered, and Half Double Tassel-flowered. It was a start.

By 1880, a more modern grouping had emerged in England: Section I, incurved exhibition varieties; Section II, very large-flowering varieties; Section III, anemone-flowered; Section IV, Japanese; Section V, anemone-flowered pompons; Section VI, pompons; and Section VII, early flowering (outdoors).

SIGNIFICANT BRITISH FIGURES

Some of the people whose stories follow were fairly prominent in their day, and there is considerable information about them. For those who are more obscure, the little information that is available has had to be gleaned indirectly. The same will be seen as the story moves from the United Kingdom to France and later to the United States.

The criterion I used to keep the list within bounds was a somewhat arbitrary limit of eight cultivars per breeder. There are records of more than 150 British men raising chrysanthemums seriously before 1900. If a breeder had introduced eight or more cultivars, I included his name. I only ignored this self-imposed rule in a few instances.

Facing Page:
Chrysanthemum cultivars: collage assembled by Yves Desjardins, science editor, *Chronica Horticulturae*
List of cultivars:
National Chrysanthemum Society bloom classification. Class 1: 'Mt Shasta', best Irregular Incurve bloom, grown by David Curtis. Class 2: 'Apricot Courtier', best Reflex bloom, grown by David Eigenbrode. Class 3: 'Golden Gate', best Regular Incurve bloom, grown by Normandie Atkins. Class 4: 'Peacock', best Decorative bloom, grown by Ron and Georgene Hedin. Class 5: 'St Tropez', best Intermediate Incurve bloom, grown by David Curtis. Class 6: 'Kelvin Mandarin', Pompon bloom, grown by Ed Mascali. Class 7: 'Peggy Stevens', best Single and Semi-Double bloom, grown by David Curtis. Class 8: 'Seatons Ruby', best Anemone bloom, grown by David Curtis. Class 9: 'Kimie', best Spoon bloom, grown by David Eigenbrode. Class 10: 'Delistar', best Quill bloom, grown by David Curtis. Class 11: 'Senkyo Kenshin', best Spider bloom, grown by David Curtis. Class 12: 'Cisco', best Brush and Thistle bloom, grown by Dorrie McDonald. Class 13: 'Lone Star', best Exotic bloom, grown by Jerry Donahue. Classes 1–12: Photographer: Todd Brethauer, Old Dominion Chrysanthemum Society. Class 13: Photographer Ralph Parks (deceased), Delaware Valley Chrysanthemum Society.
Reproduced by permission of Chronica Horticulturae

William Bull, Sr. (1828–1902), Chelsea, London

William Bull, the "new plant merchant," acquired a portion of John Weeks's nursery in Chelsea in 1861. In 1863, he leased additional space from Weeks. He purchased the nursery outright in 1874, changing the name to Bull's Establishment for New and Rare Plants. By 1878, he had become well known for introducing numerous new plants. He specialized in greenhouse plants, particularly orchids, and in pelargoniums, fuchsias, and verbenas. 'Chelsea Gem', a pelargonium he introduced in 1880, is still grown.

When Bull died (c. 1902), he had just over three acres with greenhouses. He left his business to his sons William and Edward: William Junior died in 1913, but Edward continued the business until 1920. William Bull had served on the RHS Floral Committee.

Henry Cannell (1833–1914), Swanley, Kent

Henry Cannell began his career as a jobbing gardener but was very ambitious and eventually built up a large nursery and floral business before succumbing to financial problems and ending up in bankruptcy. At its peak, the firm was well known both nationally and internationally, but the death of three of his four children had the inevitable impact on his life and ability to function. He was devastated.

In 1897 he sent specimens of his chrysanthemums to the trials at Cornell University, in New York State, together with another English breeder, Robert Owen of Maidenhead (c. 1839–1897). Cannell was very interested in many types of flowers, and was known for fuchsias, pelargoniums, and verbenas, but his most profound interest was in the chrysanthemum. He recalled that he had first seen the flowers as a child and never forgot the impression they made on him.

Cannell built his nurseries near railway lines, allowing him to send his flowers to market quickly. He understood the advantages of mail order and may have been the first nurseryman in Britain to use it. His chrysanthemums won prizes and medals at many shows. Cannell served on the Floral Committee of the Royal Horticultural Society with William Bull.

W. Clibran and Sons, Altrincham and Manchester

William Clibran had a very substantial business. He owned Oldfield Nurseries at 10 and 12 Market Street, Altrincham, Cheshire, as well as another large nursery in Manchester. Later, the firm opened more branches in other Lancashire towns, and had large seed warehouses too. At one point, Clibran employed more than 250 men. It is not surprising that the employees joined a union, and when 25 of them were dismissed in 1914, the rest struck. No doubt the onset of World War I made much of this labor dispute irrelevant. Almost no men remained to work in civilian businesses.

Clibran's lasted for more than seventy-five years, still being in business in 1960. Their earliest catalogue dates from 1881. In November 1900, the firm successfully "displayed single-flowered chrysanthemums" at the National Chrysanthemum Show. Clibran's flowers were exhibited at shows for years. At their peak, the Clibrans received a royal warrant to supply flowers to King Edward VII.

Robert Forster's name is associated with W. Clibran & Sons, although he lived and worked in Surrey. He made his living as the superintendent of the cemetery in Nunhead. It is hardly surprising that he was very active in growing flowers. He had ample space in which to experiment.

The information about his activities comes from reports in the gardening magazines of the late nineteenth and early twentieth centuries: *The Garden* and *Gardening World*. The redoubtable William Robinson edited or wrote most of the material. In volume 60 of *The Garden* (1900), Robinson reported that "W. Clibran & Sons, Altrincham, staged a collection of cut singles . . . [and] Mr. Robert Forster, Nunhead Cemetery, SE, secured a Silver Gilt Medal" for his contributions.

Robinson reported again in volume 62 of *The Garden* (1902) that "Messrs. Clibran and Forster won a silver gilt medal in a show on Monday December 27, 1902, at the National Chrysanthemum Society." The prizewinning cultivar was 'Nemasket', and the "Chrysanthemums were arranged in tall glass centre pieces." At this event, W. Clibran & Sons of Altrincham again staged "a collection of cut singles," and again, "Mr. Robert Forster, Nunhead Cemetery, SE, secured a Silver Gilt Medal."

John Freestone, Norfolk

Freestone was a very early raiser of chrysanthemums in Norfolk. Frederick Burbidge referred to him in his 1884 book on chrysanthemums. "Mr. Short and Mr. Freestone, about the year 1835, showed 'Nonpareil' and 'Norfolk Hero' at the first public Chrysanthemum show for cut blooms at Stoke Newington." In all, Freestone seems to have produced nine cultivars of chrysanthemum, and was said to be the first Englishman to raise chrysanthemums from seed, an extremely difficult thing to do.

W. J. Godfrey, Exmouth, Devonshire

Godfrey ran the Exmouth Nurseries in Exmouth, Devonshire, but also participated in some of the shows with the Devonshire branch of Veitch. Old John Veitch had started his English nursery in Exeter before branching out to the smart trade in London.

Godfrey was very industrious. Robinson reported in 1894 that he showed 'Miss Dorothy Shea', 'Charles Blick', 'Duchess of Devonshire', 'Lizzie Cartledge', and 'Aureole Virginale', among other chrysanthemum cultivars. The name 'Charles Blick' may have honored another very active nurseryman, Charles Blick, who owned Warren Nursery in Hayes. Blick won a silver Banksian Medal for his carnations but was also interested in chrysanthemums. He introduced the chrysanthemum 'Hilda Tilch' in 1910. It was a measure of his significance that he sat on the RHS Floral Committee.

In November 1895, Robinson mentioned Godfrey's chrysanthemum 'Monsieur Chas Molin', introduced from France in 1894. Later, in the same magazine, he commented on the joint display of cut chrysanthemums by Veitch and Godfrey. Godfrey showed 'Mrs. W J Godfrey', a white incurved cultivar similar to 'Mrs. Alpheus Hardy', in 1901.

He introduced many more chrysanthemum cultivars as well, such as 'Delightful', 'Yellow Boy', 'Bridesmaid', and 'Market Favourite'. Another of his cultivars, 'Bessie Godfrey', was a Japanese variety, as were 'Exmouth Crimson', 'Exmouth Rival', and 'Sensation'; the show differentiated between the Japanese varieties and others. The *Gardener's Chronicle* indicated that Godfrey grew his seedlings from his own seed. Many were rather short, not more than two and a half feet tall.

Robert Owen (1840–1897), Maidenhead

Owen owned the nursery Castle Hill at Maidenhead in Berkshire. He showed his incurved cultivar 'Lord Rosebery' at the National Chrysanthemum Society exhibition in 1893, together with about four other kinds. He also showed 'Magicienne', which won a first-class certificate, and he was the developer of 'Robert Petfield', a seedling of 'Princess of Wales'. There were also a "bronze sport from the incurved 'Robert Petfield', and 'Gold Coast', a rich, bright yellow Japanese reflexed." Robinson also liked 'Pride of Maidenhead', 'Ernest Fierens', and particularly 'Owen's Perfection'.

Owen died suddenly while working in one of his greenhouses. By the time his assistant realized something was wrong, Owen was dead.

John Salter (1798–1874), Hammersmith, London

Salter, the grandson of an English cheesemonger, ran a nursery in Versailles from 1838 until 1848. Salter was very clever and thought he would capitalize on the craze for the English garden on the Continent, but was obliged to return to England because of the Communist uprisings in Paris. He is known to have introduced more than seventy-five chrysanthemums, many of them prizewinners.

Once back in London, he set up the Versailles Nursery in Hammersmith, where his son Alfred worked with him. Salter named one of his more successful cultivars 'Alfred Salter'. Another cultivar that lasted well, 'Annie Salter', was named for his daughter. His nursery survived until 1874. When Salter retired, he sold his stock of chrysanthemums to William Bull.

Descendants of John Salter are still very active in pointing out just how remarkable a man he was. (Colin Salter was most helpful in filling in some of the details.) Salter did not start out as a professional horticulturist, but learned everything he needed to know and produced his chrysanthemums in the space of about ten years. Sometime after that, he took up pansies, and once again did stellar work. (See chapter 7.) The Canadian amateur Henry Groff followed much the same path, starting with gladioli, and then, years later, taking up iris, and seems to have been a very similar sort of person, immensely quick and energetic. (See chapter 4.)

John Salter, *carte de visite*. Photographer unknown.
Reproduced by permission of the Royal Horticultural Society

George Stevens (d. 1902), Putney, London

Stevens introduced 'Prefet Robert', "a handsome Japanese incurved flower, deep crimson in color with silvery reverse." He owned St. John's Nurseries in Putney and served on the Floral Committee of the Royal Horticultural Society.

Charles Lennox Moore Teesdale (1816–1901)

Unlike some of the men mentioned earlier, Charles Teesdale was an amateur. He was born in Guernsey to a military family just one year after the battle of Waterloo but elected not to follow the family trade. He may have been given the name Moore in admiration of a popular general.

Teesdale chose to work for the Post Office in London, gradually rising in rank. (His career mirrored that of the much better known Anthony Trollope. As part of his duties for the Post Office, Trollope invented the iconic red pillar box to post letters safely.)

Subsequently Teesdale retired to Herne, a very select part of the small Sussex town of Worthing on the south coast. There he was a justice of the peace and a magistrate. The family could afford this pleasant place because his wife had some money. (Alert readers will remember that Ernest, in Oscar Wilde's *The Importance of Being Earnest*, was found in a bag in the railway station at Worthing.)

Growing chrysanthemums was an avocation for Teesdale. He left more than sixteen new cultivars. His retirement to Herne coincided with the advent of many new greenhouses built to supply fresh fruit and vegetables to London and the other big towns. This development may have sparked his interest in flowers.

Veitch and Sons, Exeter and London

The Veitch family ran a distinguished nursery for four generations, starting in Exeter and then moving to London. They were known for very bold business moves, such as sending private collectors to many parts of the world, and offering some of the rarest and most remarkable plants on earth. In Exeter, Veitch employed John Dominy as a hybridizer. In 1861 he was the first person to introduce a hybrid orchid, an astonishing feat at the time. Introducing new chrysanthemums was part of their background activity for Veitch and Sons, not their main thrust, but as in everything else they did it very well. James Morton, the American author of a very useful book about the early chrysanthemum, wrote, "In 1881, Messrs. Veitch & Sons of London imported from Japan six new sorts, called Ben d'Or, Comte de Germiny, Duchess of Connaught, Thunberg, and others, all of which are well known."

William Wells (1848–1916), Redhill, Surrey

Wells took over Goacher's nursery at Merstham, Surrey. He published a book about the finer details of raising chrysanthemums in 1898. The book was clearly successful, for it went through several editions.

Channel Islands

The islands of Jersey and Guernsey lie closer to the French coast than the English, but are considered to be part of the United Kingdom as a result of ancient wars and battles between England and France. The islands' prosperity stems in part from their proximity to the Gulf Stream. This provides the islands with a better climate for growing crops than much of the rest of Britain. Certain commercially important flowers blossom several weeks before those on the mainland, and the Channel Islands' principal cash crop, tomatoes, ripens earlier too.

Daffodils, violets, and some bulbous plants have been associated with Jersey for a long time, but there has also been a small contingent of chrysanthemum raisers. Morton reported that M. Emile Lebois, an amateur in Paris, grew more than five hundred improved seedlings in 1836 but took advantage of the warmer climate in Jersey to "bulk up" a better crop first. He sold them to Chandlers of Vauxhall. Many of Lebois's varieties remained in commerce for a long time. Lebois was Marc Bernet's nephew by marriage (see below), and his primary nursery was in Ivry, near Paris.

Lebois was the most successful of the early breeders in Jersey during the 1840s, but he was not the first. A local baker whose name has not survived grew his plants against a wall behind his oven, protecting them from cold damage. Major Carey, a man named Clarke, James Davis, James Dawnton,

Thomas Pethers, and Charles Smith, all from Guernsey, raised new varieties for a while, but then the interest died down. Some of these men are discussed below. Davis introduced 'Prince Alfred', 'Prince of Wales', and 'Princess of Wales'. Dawnton introduced 'Elaine' and the 'Fair Maid of Guernsey'. The latter were all in Guernsey. The first exhibition of chrysanthemums was held in the islands in 1865.

BREEDERS

Major Carey, Guernsey

Another amateur, a Major Carey, worked in the Channel Islands and introduced these cultivars: 'Hackney Homes', 'Beaumont', 'Yokohama Orange', 'Victoria' (1882), 'The Czar', (syn. 'Peter the Great'), 'The Khedive', 'Sir Isaac Brock', 'Sarnia Glory', 'Red Gauntlet', and 'Mrs. C. Carey'.

The cultivar 'Lady Carey' was possibly introduced by the nurseryman James Davis in Guernsey or by Norman Davis in Sussex. Major Carey's first name is unclear. An English expert, Brian Young, told me he has narrowed down the possibilities to two men: Major de Vic Carey (1866–1904) or Major Charles Le Mesurier Carey, who died in 1905.

Thomas Pethers (b. 1821), Guernsey

Pethers worked in Guernsey and bred a lot of seedlings. John Salter bought plants from him and developed many fine varieties. Pethers introduced 'Mrs. Pethers', 'Mrs. Huffington', and 'Sir Stafford Carey', among others. (Sir Peter Stafford Carey [1803–1886] was the bailiff of Guernsey from 1845 to 1883.) Pethers traveled to South Africa for a time but did not resume breeding chrysanthemums when he returned.

Charles Smith (d. 1921), Guernsey

Charles Smith had the largest nursery in Guernsey in the nineteenth century: Charles Smith and Son, Caledonian Nurseries. He introduced several new camellias as well as the magnolia cultivar 'Goliath'. He was also very active with chrysanthemums.

Nathan Smith and Son

Nathan Smith also had a nursery in the Channel Islands. He introduced the chrysanthemum cultivars 'Mrs. E. Miles' and 'Mrs. Haliburton', among many others. As far as is known, he was not related to Charles Smith on Guernsey.

FRANCE

One of the reasons the chrysanthemum did so well in France was the warmer climate, particularly in the southwest. After the auspicious start in Marseilles, a warm Mediterranean port, the nearby Toulouse became the center for chrysanthemum development. Growers had had similar experience with roses in the semitropical Midi, as the South of France is known. Gardeners in this region did not have to contend with the cold damp of the English autumn and winter.

In 1891, a statistically minded staff writer at *Revue horticole* had the happy idea of counting how many cultivars each of the English and French chrysanthemum breeders had introduced by that time. Simon Délaux of Toulouse led with 431 cultivars, then came Auguste de Reydellet, 229, and Louis Lacroix, 202. Fourth place was taken by the Englishman Smith, with 136. The author concluded by saying that in the aggregate the top three men had introduced more cultivars than all the Englishmen combined, so there!

James Morton wondered why there was no chrysanthemum society in France at the end of the nineteenth century. He asked Victor Lemoine about it and received the following reply, dated July 9, 1890:

> We have no chrysanthemum society in France, but the numerous horticultural societies in our country are much interested in chrysanthemums, and nearly every one has a chrysanthemum show at the proper season. Pot-grown plants are generally exhibited; cut flowers in small quantities only. Here we do not grow the specimens for exhibition, as the practice is in England and America.
>
> We do not care for the enormous flowers that English florists obtain, or huge plants with only a few blooms upon them. Here the plants are treated to give the largest number of blooms, and in the most natural way. New varieties of chrysanthemums are not very largely produced in France, except in the southern portions. Here in Nancy we have a severe climate,

and it is nearly impossible to get seeds of the double varieties. Personally, we have sent out some good novelties, but the seed that yielded them was not our own. There is no country where there is so large a quantity of novelties raised annually as in France. For instance, this year, Simon Delaux, of Toulouse, offers 24 new varieties of his own production; M. de Reydellet, of Valence, 18 novelties; M. Louis Lacroix, [of Paris,] 25 varieties; M. Rozain Boucharlat, of Lyons, 14 novelties; M. Host, of Lyons, 7 novelties; M. Santel, of Salon, near Marseilles, 12 novelties; besides a number raised by Etienne Lacroix, M. Bernard, Pertuzes and Audriguier, of Toulouse, and others. Over two hundred novelties are annually produced in the south of France, principally of the Japanese and Chinese forms.

Charles Baltet (d. 1907), Troyes

Charles Baltet's father had founded a very successful nursery in Troyes. A nurseryman and horticulturist, Charles *fils* wrote *The Art of Grafting*. He and his father introduced new cultivars of many flowers widely in circulation, among them chrysanthemums. His obituary appeared in *Revue horticole* in 1908 (p. 567).

Captain Marc Bernet (1775–1855), Toulouse

Captain Marc Bernet occupies a special place in this part of the story. He had been born in Toulouse and retired there after a career in the French Army. Bernet was the first European to collect chrysanthemum seed successfully. This gave him the idea of creating new varieties. In 1827, he introduced the handsome violet-colored 'Grand Napoleon'.

Bernet handed over many of the daily tasks to his seventeen-year-old gardener, Dominique Pertuzès, and continued to introduce new varieties for many years. Eventually Pertuzès went into business for himself. Alas, he and his son François both later competed with Captain Bernet, as did many others in Toulouse.

At first, Bernet only had about thirty seeds, but in the later 1830s and the 1840s he could plant as many as three hundred seeds. He was ruthless in selecting strong and reliable seedlings from his crosses. The names of some of his early cultivars were recorded: 'Rose Croix', 'Duc d'Albuféra', 'Annibal', 'Maréchal Maison', 'Reine Blanche', 'George Sand' (a little daring

Captain Marc Bernet. *Reproduced by permission of* Chronica Horticulturae

for a provincial captain), 'Baronne de Staël', 'Princesse Pauline', and about twenty others.

Bernet became rather puffed up over his success but can be forgiven, as it was a triumphant achievement. He also had every reason to be annoyed by other people passing off his flowers as their own and selling them to make money. At first he had been very generous and shared his results with many horticulturists, but because there was no law of copyright at the time and no one even thought about patenting living things, there was no official way to safeguard his plants. He had to protect his ideas and work himself. One person he trusted was his niece's husband, Emile Lebois, and only Emile was allowed to grow his new cultivars. Bernet sometimes wrote under the pseudonym "Dr. Clos." He was also known as the Chevalier Bernet, for the decoration he received.

Emile Lebois later left Toulouse, moved to Paris, and started a new chrysanthemum business. He shared the bounty with three other upright men: Auguste Miellez in Lille, John Salter in Versailles, and Philippe Pelé in Paris. In 1854, Pelé introduced the first successful line of dwarf pompon chrysanthemums.

Captain Pierre-Louis Blancard.
Reproduced by permission of Chronica Horticulturae

Pierre-Louis Blancard (1741–1826), Marseilles

Captain Blancard came from an old Marseilles family. He was born and died in the city. Blancard went to sea very young with his father and despite a rather attenuated education became interested in commerce and geography. In 1813, he wrote a brief treatise, *Manual on the Commerce of the Indies and China,* which was published for the first time by the Geographical Society of Marseilles in 1910, almost a century later. After retirement he joined the Agricultural Council in Marseilles.

Once he obtained his captain's certificate, the merchant family of Audibert employed him in their shipping line. His first trading voyage for the Audiberts was in 1770 to the Île-de-France (now Mauritius). The ship left from the port of Brest and returned there two years later. No one was surprised by 22 sailors dying of scurvy while they were at sea, although English sailors were already being protected by the use of lime juice.

On the fifth voyage, leaving Marseilles in 1787 and lasting almost three years, he sailed to the Île Bourbon in the Indian Ocean (now Réunion), Bombay (now Mumbai), the Maldives, Sumatra, and Singapore. His next move was to go to China. He found a pilot on the island of Wampoa who took the ship to Canton. Almost the first thing he did there was to buy half a dozen chrysanthemum plants.

The return journey took fifteen months, and only three of the plants survived. He acclimatized the survivors in his garden in Aubagne, Marseilles. One of them, which was tall and purple, became known afterward as 'Old Purple'.

Blancard made a similar trip two years later, in 1791. The French Revolution had broken out while he was away on the fifth voyage, and perhaps he felt happier being at sea a little longer. For all its hazards, the sea was a bit more secure than the unpredictable events of the revolution.

Blancard's voyages were documented in considerable detail, and the records are in the archives in Marseilles. Many years later, his granddaughters, who lived in England, were found to be in extreme penury. Charitable members of the National Chrysanthemum Society of England took up a collection to help alleviate their distress. The city of Marseilles named one of its main streets Promenade Blancard, and a small alleyway is also named for him in the city; his house has had a plaque on it since 1938.

François Bonamy et Frères, Toulouse

The Bonamy brothers were landscapers and nurserymen in the Place Dupuy, Toulouse. In about 1850 they developed the miniature anemone class of chrysanthemum: 'Eucharis', 'Medee', and 'Thisbe' are examples.

A. Bonnefous, Moissac

Monsieur Bonnefous was a gardener at the Jardin de Landerose in Moissac. About fifty cultivars of chrysanthemum are attributed to him.

Laurent Boucharlat (1806–1893), Lyon

Boucharlat founded his business in 1833 and showed his mettle by winning a medal at the Lyon Flower Exposition in 1838, though it is not clear which flower he submitted. He experimented with miniature chrysanthemums and introduced several of the pompon variety. 'Mme Custex Desgranges', a very important early ("hâtive") white, appeared in 1873. It was the basis of future "corbeilles automnales." In addition to chrysanthemums, Boucharlat was noted for his pelargoniums and petunias.

Had the Franco-Prussian War not turned France upside down after 1870, Boucharlat would have been awarded the Legion of Honor, but this did not happen. There is no question of his importance in the French horticultural community. They mourned his death very sincerely.

Laurent's younger brother was Jean Marie Boucharlat (1818–1903). He too worked in Lyon.

Ernest Calvat (1858–1910), Grenoble

The Calvat family was prominent in Grenoble. M. Calvat's father, also Ernest Calvat, served as mayor from 1871 to 1873. The future *chrysanthéemiste* was born in 1858 and christened Jean Marie Ernest but was always known solely as Ernest. Calvat was definitely an amateur. He owned a successful glove factory, which allowed him to live well and devote himself to his flowers. Chrysanthemums were his principal interest. At one time he was president of the Horticultural Society of the Dauphinée, indicating his position in the horticultural hierarchy of his time. Calvat introduced dozens of significant new cultivars. His accomplishments were recorded in an obituary in *Revue horticole*.

Alfred Chantrier, Bayonne

Chantrier was gardener to a Monsieur Bocher and was active between 1885 and 1896. He ultimately introduced forty-five new cultivars, of which 'Duchesse d'Orléans' and 'Candeur des Pyrénées' were notable.

Anatole Cordonnier (1842–1920), Bailleul

Cordonnier wrote a short book and several articles in *Revue horticole*. He was extremely enthusiastic about chrysanthemums with very large blossoms. From the record, he seems to have bred only ten new cultivars, but his fascination with these flowers suggests that he may have bred others of which no records have been retained.

Simon Délaux (1840–1902), Toulouse

Délaux lived and worked in St. Martin de la Touche, near Toulouse. He was a major figure in the early chrysanthemum world and used intentional cross-fertilization most effectively. He introduced more new cultivars than anyone else. The great English horticultural writer William Robinson wrote Délaux's obituary for *The Garden* in the most glowing terms. Robinson lamented the fact that though they had a very valuable correspondence for many years, he had never met Délaux in person.

He noted that Délaux's work was held in very high esteem among English gardeners, as Calvat's was to do later. Délaux's work on the Japanese imports was of especial interest. Among his best-known cultivars were 'Mme Berthier Rendatler', 'M Astory', and 'Japonais'. Délaux also worked with the early flowering types, which found many enthusiastic growers in England. There are numerous references to his new cultivars in the gardening magazines of the 1890s, listing extraordinarily good flowers. The fluffy incurved 'Comte F. Lurani', which blooms in October, received particular attention. It stood the test of being grown for several years, performing just as well three years after its introduction.

Monsieur Hoste (1820 [1823?]–1894), Monplaisir

No first name has come down to us. Hoste was born in Gand (also known as Ghent), Belgium, and moved to France. He established his nursery in rue de Dahlia in Monplaisir, naming the street himself. Hoste was among the earliest to cultivate chrysanthemums in France, and his work served to introduce the flower to the greater public. He seems to have retired and handed over his business to André Charmet, a fellow Belgian. Some of his cultivars are 'Ami Jules Chrétien' (c. 1890) and 'Catros Gerarde' (c. 1895).

Etienne Lacroix

Etienne Lacroix was a professional nurseryman whose most popular cultivars were 'Parasol', 'Mlle. Lacroix', 'Flocon de Neige', 'Jeanne d'Arc', and 'Fabias de Mediana'.

Dr. Louis Lacroix, rue Lancefoc, Toulouse

Louis Lacroix was an amateur, making his living from a fireworks business. One of his chrysanthemum cultivars was 'Viviand Morel'. He was said to have nine hundred varieties of plants in his garden, but these could not all have been chrysanthemums.

Emile Lebois, Livry, near Paris

Lebois worked in Livry, near Paris. As mentioned earlier, he was married to Captain Bernet's niece. After his death, his widow maintained his work. At her request, a small committee from the horticultural society of Haute-Garonne in Toulouse visited her in 1873 to appraise how well she was carrying on her late husband's business. They were impressed by her work in developing seven new cultivars.

One of the committee members was a Monsieur Pertuzès, the son of the young Pertuzès who had worked for Captain Bernet. The committee recommended Madame Lebois for an honorable mention in the annals of the society, noting that she burnished the firm's reputation in a most worthy manner. Reporting on other visits, Pertuzès and his colleague Monsieur Marrouch complained about the confusion in chrysanthemum nomenclature that existed and suggested methods to combat this.

Louis-Jules Lemaire (1859–1925), 26 rue Friant, Paris

Louis-Jules Lemaire was Philippe Pelé's grandson (see below). The last professional nurseryman to grow his own plants in Paris, Lemaire was known to be a master hybridizer. His two sons Louis and Paul worked in Bagneux but later moved to Saint Jean-de-Braye, near Orléans, in 1949. His granddaughter Paulette Lemaire collected as much information about her family and its work as she could, and she developed the Conservatoire National du Chrysanthème in Saint Jean-de-Braye to commemorate them all.

Victor Lemoine, rue du Montet 134, Nancy

Surprisingly little has been written about this most amazing of all the heroic hybridizers of the nineteenth century. Without his work, it is likely that garden centers as we know them might never have developed or would have come much later. An enormous proportion of the standard annuals, perennials, and flowering shrubs in commerce came from his nursery. If there is one plant with which his name is forever associated it is lilac, *Syringa*. Lemoine did not pay a great deal of attention to the chrysanthemum, but nevertheless introduced a respectable number of very good new cultivars.

Auguste Miellez, Esquermes les Lilles

Miellez was important because he worked quietly with plant crossing in the early 1830s while it still was considered to be a suspect activity and somewhat impious. Lemoine knew about him as a very young man and spent several months in his nursery learning the techniques. Miellez is perhaps best known for his roses.

Auguste Nonin (1856–1956), route de Paris, Châtillon-sur-Bagneux

Auguste Nonin inherited his father's nursery in Châtillon at the early age of fifteen. His father, Emile Nonin, was killed in his own garden by a sentry during the Franco-Prussian War. His mother then ran the business and continued to rear her five children alone. Once Auguste married in 1880, he took over from her and started on his remarkable career as a developer of new plants.

Nonin was very observant and understood the conditions in which he had to work, the markets, and the world of competitions. As a young man he traveled widely, particularly to England, where his work on chrysanthemums was greatly appreciated. He won a Certificate of Merit at the London exhibition in 1905 for 'Perle Chatillonnaise', a large creamy-white blossom tinged with pink. 'Coquette de Chatillon', 'Chatillon', 'Sarah Bernhardt', 'Président Truffaut', 'Président Loubert', 'Raymond Poincaré', and 'William Turner' were among his most successful introductions. He was also decorated by the French government and served on many juries at flower shows. This record of success led to him being elected a vice president of the Société Nationale d'Horticulture Français.

In spite of all these achievements with chrysanthemums, Nonin is remembered today as a distinguished rosarian. He always refused to be pigeonholed as a specialist, but not unlike Victor Lemoine worked with a broad variety of flowers. The Nonin catalogues indicate this breadth of interest.

In 1912, he offered the very large-flowered chrysanthemums bred by Ernest Calvat of Grenoble. In 1913, his catalogue contained more than twenty chrysanthemum cultivars, of which eight were new and came into flower very early in the season, "très précoces." In that same catalogue, he listed pansies, dahlias, geraniums, fuchsias, cannas, begonias, and new roses, many of which he had bred himself. Another series, perhaps in 1914, listed early-flowering dwarf chrysanthemums.

Nonin's son Henri inherited the nursery in his turn, and continued to improve it until 1945. Henri was also a fine rosarian. The business finally closed in 1960, most probably for the same reasons the Lemoine nursery closed in Nancy at the same time. They had survived two punishing world wars but could not compete in the postwar environment. Too many external forces were making it more and more difficult to sustain a nursery in France. The Netherlands and Belgium began exporting the same flowers grown in France far earlier in the season. They forced them into bloom in huge greenhouses.

André-Philippe Pelé, Paris

Another early enthusiast was the Paris nurseryman André-Philippe Pelé. He raised his seeds in the south and was extremely thoughtful about which ones he selected for further study. Pierre Coindre of Avignon had bred the first early-blooming chrysanthemum in 1850. In 1855, Pelé exhibited his own series at the Société Nationale d'Horticulture show.

Dominique Pertuzès, rue des Chalets, Toulouse

Pertuzès started out as Bernet's gardener at the age of seventeen but later began his own business. Pertuzès stole Bernet's work and competed with him rather unscrupulously.

Alexandre de Reydellet, Valence

De Reydellet was the stationmaster at Bourg-les-Valence, Drôme, and chrysanthemums were his hobby. Some biographical information is available on de Reydellet (in *Lyon horticole* [1905] and *Annuaire de la Société Nationale d'Horticulture de France* [1899]), though where and when he was born remain a mystery. According to the *Revue horticole* (1905),

> Alexandre de Reydellet died in October 1905. He was an amateur horticulturist in Bourg-les-Valence (Drôme). De Reydellet was a member of the Association Horticole Lyonnaise since 1891, and member of the SNHF since 1886. He was one of the founders of the Société Française des Chrysanthémistes. He was one of the first to sow chrysanthemums at a time when few seem to have observed that chrysanthemums gave seeds. He began about 1875 or 1877. He gave his first cultivars to Boucharlat the elder (Lyon) in 1882. Then he started to sell them himself. He received a lot of awards. The first medal of honour of the Société Française des Chrysanthémistes was for him. He was made Chevalier du Mérite Agricole.

One of de Reydellet's earliest cultivars was 'La Triomphante' in 1877.

Joseph Rozain-Boucharlat (1849–1917), Lyon

Joseph was Laurent Boucharlat's nephew. He had a very distinguished career, including founding and becoming president of the Société Française des Chrysanthémistes (an organization based on the English society) as well as vice president and councilor of the Société d'Horticulture Pratique du Rhône. He had studied in England and, amazingly for a Frenchman, admired English ways. Rozain-Boucharlat also worked with fuchsias, dahlias, and pelargoniums. His son Benoit (1886–1943) worked at Cuire-lès-Lyon.

Vilmorin-Andrieux, Paris

The seed house Vilmorin-Andrieux has been at the same address in Paris for more than two hundred years. It might be said to have become legendary. Philippe-Victoire Vilmorin (1746–1804), a physician with a keen interest in plants, founded the firm in 1775 after marrying the daughter of Pierre Andrieux, a seedsman and botanist in the quai de la Mégisserie. Andrieux's wife, Claude Geoffroy, was the expert. Together they became the suppliers of seed to Louis XV, a huge advantage in those days. What happened at court set the standard for everyone else. The firm prospered for six generations, but about thirty-five years ago it was sold to a large conglomerate and is now part of Groupe Limagrain.

At the outset, Vilmorin-Andrieux concentrated on agricultural seed. Pierre-Victoire's sons and grandsons developed important strains of sugar beets and carrots. These were very sensible if somewhat unglamorous business decisions and allowed Philippe-André (1776–1862) to move his family into an elegant chateau at Verrières, a former hunting lodge of Louis XIV, in 1815. They and their descendants transformed the park, designed by Le Nôtre, into an outstanding arboretum. It is just possible that this move may have been eased by the flight of its former aristocratic owners during the revolution. Vilmorin must have played his cards very cleverly to avoid being executed, in light of his association with the royal house.

Louis Vilmorin (1816–1860) did his work on the sugar beet at Verrières. His son Henry, an authority on the genetics of wheat, understood plant genetics very early and contributed to the advance of that science. The Vilmorins also maintained a key collection of potatoes. The company added ornamental plants very early and became known for its roses, introducing new varieties for many years. One or another of the Vilmorin brothers was always in demand as a judge or a speaker at floral society events. They also won prizes at chrysanthemum shows.

In the twentieth century, the sons continued to manage the firm, but Louise, the only girl, rebelled, becoming an avant-garde poet and novelist. She married and moved to the United States. In the complex shifts among large commercial horticultural enterprises over two hundred years, Vilmorin-Andrieux is one of the few companies that remained in business into the recent past. Its founders would not recognize it now, but adapting to change and moving forward are both qualities of successful firms.

MINOR FIGURES, FRANCE

Dr. Audiguier

Dr. Audiguier bred 'Soleil Levant', a rather memorable cultivar.

Monsieur Bernard, Toulouse

M. Bernard is another of those enigmatic breeders who left no other trace besides his cultivar 'Gloria Rayonnante'.

André Charmet (1823–1897), Lyon

Charmet was born in Ghent, Belgium. He took over Hoste's lucrative business in Lyon and continued to breed chrysanthemums.

Pierre Crozy l'aîné (1831–1903), Lyon

Pierre started a nursery at 206 Grande-rue de la Guillotiere, Lyon. His son Michel Crozy (1868–1906) took over his father's nursery at his death when Pierre died. The late Thomas Brown, who reconstructed authentic historical landscapes, using only plants available during the relevant epoch, listed the dates of this establishment as 1870 to 1908.

Jean Heraud, Pont d'Avignon

Heraud was head gardener at the Villa Brimborion, Pont d'Avignon, in Provence. Only the name has endured, but he left a legacy of new chrysanthemums.

Marquis de Pins, Montbrun near Toulouse, Gers

The marquis owned a chateau at Montbrun near Toulouse, Gers, and devoted himself to breeding new chrysanthemums. The name de Pins appeared quite frequently in the horticultural literature of the day.

UNITED STATES

The chrysanthemum appeared very quickly in the United States after its arrival in Europe. Many other flowers had a similar trajectory. The first known hybrid chrysanthemum cultivar in the United States, 'William Penn', was exhibited by Robert Kilvington of Philadelphia at the annual meeting of the Pennsylvania Horticultural Society in 1841. At that stage, all chrysanthemums were still grown outdoors, and both professional and amateur breeders worked with garden chrysanthemums. After about 1850, their culture was transferred to greenhouses. Amateur breeders took up the greenhouse flower very soon after this transition.

Some of the most notable were Charles Totty of Madison, New Jersey; Eugene H. Mitchell of the Dreer Company in Philadelphia; and Elmer Smith of Adrian, Michigan. Smith began his work at the end of the nineteenth century and by 1923 had introduced 445 cultivars.

Charles Mason Hovey wrote in his 1846 Massachusetts horticultural magazine that "few plants afford more gratification than a good collection of chrysanthemums." Soon after, the Pennsylvania Horticultural Society held a show in 1846 promoting it as "the coming flower." Within the next twenty years, the flower consolidated its hold on gardeners' imaginations. By about 1865, chrysanthemum shows were springing up in many towns and growers competed for prizes and medals just as in the old country. The venerable Pennsylvania and Massachusetts Horticultural Societies were in the forefront of this movement.

As this ferment continued, the introduction of the exquisite white Japanese variety, 'Mrs. Alpheus Hardy', took the Western world by storm. A Japanese student at Harvard wanted to please his mentor, Professor Hardy, and arranged for a specimen of this flower to be sent to the professor in 1891, naming it after the teacher's wife. It was a precursor of the "spider" type, with exquisitely curved and fluffy petals.

As Morton observed in his 1891 book,

> The chrysanthemum has been exhibited at the shows of the Massachusetts Horticultural Society in Boston since 1830. The list of varieties exhibited at that time was as follows: 'Quilled Flame', 'Curled Lilac', 'Tasseled White', 'Golden Lotus', 'Large Lilac', 'Changeable Buff', 'Paper White', 'Crimson',

'Mrs. Alpheus Hardy' chrysanthemum
From Garden and Forest, *February 29, 1888*

'Pink', 'Lilac', 'White', 'Semi-quilled White', 'Parks', 'Small Yellow', 'Golden Yellow', 'Quilled Lilac', and 'Quilled White', these being exhibited by Robt. L. Emmons of Boston, then recording Secretary of the Society, and Nathaniel Davenport. The plants were spoken of as grown in the open ground and evidence is given that that the number of varieties at this period was very small. They were exhibited on the 20th of November, and reported in the New England Farmer of November 26th, 1830.

MAJOR FIGURES

John Lewis Childs (1856–1921)

Childs started his career in the seed business in 1874, when he went to work for a nurseryman in Queens, New York, C. L. Allen. In 1895, Childs won Best in seven varieties, plus a cultural certificate. The exhibit contained 'Ivory', 'President Smith', and 'Eugene Dailledouze'. In 1898 he exhibited 'White Maud Dean' to great acclaim at the Chrysanthemum Society of New York. Childs was elected state senator twice, 1894 and 1896, but failed to get elected to the U.S. Congress. He founded the John Lewis Childs Seed Company in Floral Park, New York; his nursery was so large that the town was named Floral Park for it. After his death, his widow continued to operate the business for a time, but it finally closed down in the 1930s because of the Great Depression.

Childs bought many of his seedlings from Thomas Pethers and other nurserymen in Guernsey. The 'Fair Maid of Guernsey' (mentioned above in the British section) is one of the best known.

John Condon (1844–1902), 734 Fifth Avenue, Brooklyn, New York

Condon appears to have been a very enterprising businessman and good at promoting himself. This was the background against which he bred chrysanthemums in his greenhouses along Fort Hamilton Parkway in Brooklyn. He ran a successful florist shop on Fifth Avenue and played a role in Brooklyn's civic life. He was an active Democrat and managed to get his name in the local newspaper, the *Brooklyn Eagle*, quite frequently.

The *Brooklyn Eagle* indicates that Condon held yearly chrysanthemum shows at his Brooklyn greenhouses. He also entered the major chrysanthemum shows in town and received commendations. Condon's business was near the Greenwood Cemetery. He regularly advertised fresh flowers for funerals at low prices, as well as flowers for weddings.

C. E. Honn, the gardener for the Cornell College of Agriculture, admired his work. He sent Condon a letter in November 1900 listing the cultivars he held in particular esteem: 'Columbia', a plant with a seven-inch flower (Honn said that L. H. Bailey's wife also liked this flower a great deal; Bailey was the revered dean of the college), 'Shamrock', 'Golden Ball' ("a close second to Roosevelt"), 'Brooklyn' ("one of the best pinks"), 'W. J. Bryan', 'Henry Ward Beecher', 'Clara Barton', 'Hobson', and 'Cornell'.

Condon published this letter in the *Brooklyn Eagle* to puff himself. He was mentioned in other publications too, with one of his achievements "being first for best twenty-four varieties" in 1893 at the annual Chrysanthemum Show of the New York Florists' Club in a new building at Forty-Third and Lexington. Condon was buried in the Calvary Cemetery in Queens.

Frederick Dorner and Sons, Lafayette, Indiana

Dorner was born in Germany in 1837. He immigrated to the United States to work with his brother Philip in Lafayette, Indiana, in 1855, when many ambitious young German men were crossing the Atlantic Ocean to improve their lots. It was hard for a poor man to buy land in Germany, and life remained very class ridden and stratified. Georg Ellwanger, of Ellwanger and Barry in Rochester, New York, was a prime example of how successful it was possible for a young German lad trained in gardening to become in the United States. Dorner died in 1911, leaving a legacy of beautiful new chrysanthemum cultivars.

Edwin Fewkes and Sons, Newton Highlands, Massachusetts

Morton wrote, "There is no commercial house in New England more favorably known to chrysanthemum growers than that of E. Fewkes & Sons.... At one time these gentlemen held the entire stock of 'Mrs. Alpheus Hardy', and were the first to flower and exhibit it in America; they still retain the silver medal awarded to its first bloom by the Massachusetts Horticultural Society. It was also from this bloom the first cut was made that illustrated the horticultural papers and catalogues at that time."

The first varieties offered in 1868 by Messrs. Fewkes "met with but little sale, and out of their entire collection," we are informed, the variety 'White Treveana', a small

> double white flower, was the only one that commanded even a passing attention. The house of Edwin Fewkes & Sons has steadily kept pace with the increasing interest in the chrysanthemum, and to their skill as growers and enterprise as importers we are indebted for the following excellent varieties: 'Wm. H. Lincoln', 'Kioto', 'Neesima', 'Lilian B. Bird', 'Mrs. Fottler', 'Belle Hickey', 'Emmie Ricker', 'Nippon Medusa', 'S. B. Dana', 'Marian', 'Clarence', 'Bryant', 'Emily Selinger', 'Flora', 'Nahanton', 'H. A. Gane', 'Jno. Webster', 'James F. Mann', 'Lizzie Gannon', 'Pres. Hyde', and chief of all, the far-famed 'Mrs. Alpheus Hardy'.

Peter Henderson (1822–1890), New York

Henderson, a Scottish nurseryman who settled in New York, rapidly became one of the best known and most respected of all the people who ran nurseries at the time. He was known to be a prodigiously hard worker, sitting up late at night to write his innovative series of manuals about practical gardening and horticulture for profit. He believed that any intelligent person could become a successful market gardener without undergoing a long, complicated apprenticeship. After his death, his business continued on to the middle of the twentieth century.

Henderson also bred new varieties and was involved with many different types of flowers. When Theodosia Shepherd in Ventura developed new strains of giant petunias, she sent the results to him. He blessed her work and told her she could start a whole new industry of American flower seeds in packages. His endorsement was of immense importance at the time.

In 1875, he published *Gardening for Pleasure*. He devoted a brief chapter to the chrysanthemum, describing its fairly recent ascent in popularity and the development of increasingly frequent shows. Together with succinct cultural

Peter Henderson, pioneering Scottish nurseryman in New York.
Reproduced by permission of LuEsther T. Mertz Library, New York Botanical Garden

advice, he listed the plants he thought would be best for amateurs to grow. The book went through several editions; the one I consulted is from 1888.

Early: 'Bouquet Nationale', 'Bouquet Fait', 'Elaine', 'Red Dragon', 'Gloriosum', 'J. Collins', 'Mrs. Brett', 'Mme Grame', 'Mrs. S Lyon', and 'Sonce d'Or'.

Late: 'Bend d'Or', 'Cullingford', 'Count of Germany', 'Christmas Eve', 'Fantasie', 'Fair Maid of Guernsey', 'Golden Dragon', 'James Salter', 'Lord Byron', 'Lady Slade', 'Mrs. C. L. Allen', 'Moonlight', 'Mrs. C. H. Wheeler', 'Maid of Athens', 'Talford Salter', and 'Yellow Eagle'.

The names alone show how diverse the plant material was in New York at the end of the nineteenth century. 'Mrs. C. L. Allen', for example, was named for the wife of his old employer on Long Island, where Childs began his career. There are some obviously French names here, such as 'Bouquet Nationale', 'Bouquet Fait', and 'Sonce d'Or'. 'James Salter' and 'Talford Salter' refer to James Salter, an English nurseryman who worked in France for ten years but moved back to England in 1848.

More than twenty years later, in the 1911 edition of *Practical Horticulture*, Henderson had only added one or two varieties to his list from the 1880s.

Edward Gurney Hill (1847–1933), Richmond, Indiana

The Hills primarily grew roses but were known for chrysanthemums, carnations, and other flowers too. The firm closed for good in 2007. Morton writes:

> This year [1888] Messrs. Hill & Co., of Richmond Ind., sent out of his raising 'C. A. Reeser', 'John Lane', 'Mrs. Winthrop Sargent', 'Carry Denny', 'Reward', 'Model', 'Twilight', and 'White Cap'. The colors and tints which were unknown in this flower a decade ago are now found in all of these varieties. Maroons, crimsons, rose, pink and buff have become more decided, and with such progress as this in another decade, the production of a scarlet flower is not to be despaired of by those who have done most in our favored climate to bring out the newer and formerly unknown shades.

Hill is also brought to life by Simon S. Skidelsky, a somewhat unlikely source. Skidelsky was a traveling commission merchant who branched out on his own after he began selling florists hardware for a rather frugal master in Philadelphia. In particular, he devoted a lot of time and energy to trading in carnations, and he got to know some of the more preeminent carnation breeders quite well.

He traveled back and forth across the continent by train, spending time with people like E. Gurney Hill during the workday and afterward relaxing over a drink or dinner. Occasionally, the grower was an austere teetotaler, like Richard Witterstaetter of Cincinnati (see chapter 5), but Skidelsky took that in his stride. As a consummate salesman, he only saw the good in people and endeavored to charm them to the best of his ability. It was hard to charm Witterstaetter, but he managed it.

George Hollis (1839–1911), South Weymouth, Massachusetts

Hollis left at least nine new varieties of chrysanthemum.

Robert Kilvington, Philadelphia

Robert Kilvington exhibited the first known hybrid cultivar in the United States, 'William Penn', at the annual meeting of the Pennsylvania Horticultural Society in 1841.

C. D. Kingman

C. D. Kingman won a first-class certificate of merit for his seedling chrysanthemums 'Eglantine' and 'Kildare' at the November 1890 Massachusetts Horticultural Society chrysanthemum show. In 1892 he received another certificate for 'Nemasket', a very large pure white chrysanthemum.

George W. Miller, Chicago

Miller was a florist in Chicago and supplied a wealthy citizen, John Lane, with flowers for special events. Morton wrote that with "such a name as that of John Lane at the front, we may expect much from the west in the not distant future. Mr. Lane is a retired businessman, an enthusiastic amateur in chrysanthemum culture, and treasurer of the National Chrysanthemum Society. He has extensive grounds and several greenhouses, from which his friends and neighbors reap the benefit, for his flowers are distributed with the most lavish generosity. His critical notes on varieties and culture, written in a style wholly his own, always receive great attention."

Pitcher and Manda, United States Nurseries, Short Hills, New Jersey

Morton wrote, "The enterprise . . . of Messrs. Pitcher & Manda, of Short Hills, N. J., has given much to the lovers of chrysanthemums on this continent." In 1889, the company imported from Japan and distributed 'Rohallion', 'Passaic', 'Kansas', 'Arizona', 'Ithaca', 'Raleigh', 'Jean Humphreys', and 'Mrs. Cornelius Vanderbilt'. They also developed themselves and distributed 'Bohemia', 'Indian', 'Iona', 'Iowa', 'Iroquois', 'Oneida', 'Mohawk', 'Virginia', 'Pequot', 'Minnewawa', 'Connecticut', and 'Mrs. DeWitt Smith'. Pitcher and Manda also obtained the entire stock of 'Mrs. Alpheus Hardy' from Edwin Fewkes and Son and distributed it to the public for the first time.

Elmer D. Smith and Co., Adrian, Michigan

Elmer Smith (1854–1939) bred many new cultivars. Because of his dedication to the chrysanthemum, he wrote a book about its culture that ran into at least four editions.

Thomas H. Spaulding, Orange, New Jersey

Morton stated that Spaulding would "long hold an important place among the chrysanthemum growers in America. This gentleman sent out his first seedling in 1886, and each year since then many excellent varieties of his production have been placed upon the market." In 1888, Spaulding distributed 'Geo. McClure', 'Mrs. John Pettit', 'Cloth of Gold', 'Eleanor Oakley', 'E. S. Renwick', 'Gladys Spaulding', 'Juno', 'R. E. Jennings', and others. In 1889, he produced and released 'George Atkinson', 'Commotion', 'Tusaka', 'Takaki', 'Mrs. Judge Benedict', 'We Wa', and 'Brynwood'; he also developed 'Addie Decker', 'Maria Ward', 'Garnet', 'Mrs. Thomas A. Edison', 'Jas. R. Pitcher', 'Cyclone', and 'Zenobie'. He distributed the English prize chrysanthemum, 'Mrs. S. Coleman', and his own seedling, 'Ada Spaulding'.

'Ada Spaulding', a cross between 'Puritan' and 'Mrs. Wanamaker', was awarded the National Prize in November, 1889 in Indianapolis, as well as other honors including a certificate of merit by the Pennsylvania Horticultural Society, a first premium by the New Jersey Horticultural Society, and a medal of excellence by the American Institute, in New York. Morton describes it as "of robust habit; a rich deep pink, shading in upper portion to the purest pearl white; globular in shape and neither Japanese nor Chinese in form."

Even more varieties first distributed by Spaulding, some imported from Japan and others grown in his own greenhouses, include 'G. F. Moseman',

'Mrs. T. H. Spaulding', 'Sokoto', 'Leopard', 'Mrs. J. N. Gerard', 'Pauline', 'Coronet', 'Dango Zaka', 'G. P. Rawson', and 'Peculiarity', as well as 'Miss Sue Waldron' and 'Snowdrift', grown by Mr. J. N. Gerard, of Elizabeth, New Jersey, and 'Sunset', 'Mrs. Wm. Barr', 'Alice Brown', and 'Fannie Block', grown by William Barr, of Orange, New Jersey.

Morton wrote, "There are also many other excellent varieties, either raised or disseminated by Mr. Spaulding, that chrysanthemum lovers now enjoy, several of the seedlings of Messrs. Lord, Allen and Hollis being among them."

John Thorpe (1842–1891), Pearl River, New York

Thorpe was born in England but moved to the United States in 1874. For a time he worked for the firm of W. Hallock and Sons in Queens, New York. Morton wrote in 1891,

> The progress of chrysanthemum growing in America can not well be written without mention of the firm of V. H. Hallock & Son, Queens, Long Island. To these gentlemen we owe the origin of many excellent sorts, to the number of which they are constantly adding, as is evidenced by the list of new varieties that are offered annually to the public through their catalogues. This year they offer twenty new varieties in one collection for the first time. Among the varieties which they have been instrumental in giving to the public are 'Mrs. Langtry', 'W. Falconer', 'Whirlwind', 'Pagoda', 'Sadie Martinot', 'Frank Wilcox', 'T. F. Martin', 'Moonflower', 'Mrs. Cleveland', 'F. T. McFadden', 'Mrs. Potter', 'Edwin Booth', 'Prince Kamoutska' and 'V. H. Hallock'.

While working there, John Thorpe produced some excellent varieties, and sent out his first seedlings in 1883.

In 1893, he was appointed the chief of floriculture at the Chicago World's Columbian Exposition, but subsequently settled in upstate New York, opening his own business. The influential William Robinson referred to him as "the father of the chrysanthemum in America," and Morton wrote, "The name of John Thorpe is well known to chrysanthemum lovers throughout America, as well as in England."

In his day he was a very well-known and significant figure. Clearly he had considerable organizational and administrative ability. Thorpe was known for his prizewinning dwarf varieties but also many other types of chrysanthemum. In 1888, he won the Carnegie Silver Cup for 'Mrs. Andrew Carnegie'. He served as secretary to the New York Horticultural Society, was the first president of the important Society of American Florists, and was a founding member of the Chrysanthemum Society of America (in Buffalo, New York) in 1890. Despite his renown and achievements, Thorpe became impoverished as he grew old, and died in very sad circumstances.

Charles H. Totty (1873–1939), Madison, New Jersey

Totty was born in Shropshire and immigrated to the United States when he was twenty years old. By 1896 he had settled in Madison, New Jersey. He opened a nursery in New Jersey and also ran a very successful florist shop at 4 East Fifty-Third Street in New York, right at Fifth Avenue. One thinks of Edith Wharton's *The Age of Innocence* and the hero ordering yellow roses every day for his lover, in the heyday of Fifth Avenue as the acme of social triumph. Perhaps Wharton was thinking about Totty's.

Some of his best-known chrysanthemum cultivars were the prizewinning 'Patricia Grace', 'Mrs. Henry Evans', 'White Cheifton' (perhaps THIS really should be "Chieftain"), and 'Amaterasu'. Totty also operated an active mail-order business. The New York Horticultural Society awarded him a medal in 1913 for developing a new rose, 'Shell Pink Shawyer'.

Dr. Henry Pickering Walcott (1838–1932), Cambridge, Massachusetts

Dr. Walcott was an expert on public health and represented Massachusetts on the Advisory Council of the American Public Health Association in the 1880s, as well as being a founding professor of the Harvard School of Public Health. An avid gardener, Walcott was recognized in his day as a distinguished amateur horticulturist and received many awards. Walcott also served as president of the Massachusetts Historical Society in 1886 and again in 1904. The record indicates that he bred delphiniums and many other types of flowers.

As far as is known, Walcott was the first person to raise Japanese and Chinese chrysanthemums successfully from seed in the United States. He found that if he kept everything very dry he could be successful. This signal achievement attracted very little attention at first, but subsequently the floral community recognized its importance. Dr. Walcott showed some of his seedlings in 1879.

Walcott entered 'Sport', 'Nil Desperandum', and several other varieties at the Massachusetts Horticultural Society chrysanthemum show in November 1888. He considered his best chrysanthemum cultivars to be 'R. Walcott', 'Shasta', 'Savannah', 'Wenonah', 'Monadnock', 'Semiramis', 'Alaska', 'Ramona', 'Nevada', 'Cambridge', 'Pontiac', and 'Tacomah'. Walcott's use of Native American names was a novel departure for the time. Henry Pickering Walcott was an outstanding man, good at everything he did and concerned for the public welfare.

Henry Waterer, Philadelphia, Pennsylvania

Waterer was the name of a very distinguished family of nurserymen in England, and Henry might have been a relative. Morton wrote, "In 1883, Mr. H. Waterer, of Philadelphia, brought an importation from Japan of some fifty varieties, many of which were most distinct and beautiful, which gave a new impulse to hybridizing, as from that time to the present, the new kinds that have appeared annually are almost numberless."

Waterer's Japanese imports included 'Gloriosum', 'Mrs. C. H. Wheeler', 'Marvel', 'J. Collins', 'Duchess', 'H. Waterer', 'Pres. Arthur', 'Snowstorm', 'Mrs. Geo. Bullock', and 'Mrs. Vannaman'. In 1885, Waterer distributed 'Puritan', 'Miss C. Harris', 'John M. Hughes', 'Miss Meredith', and 'Mrs. R. Mason', and in 1886, 'Wonderful', 'Robt. Crawford', 'Mrs. John Wanamaker', 'Thos. Cartledge', 'Alfred Warne', 'Mrs. Anthony Waterer' and 'Lucrece', all produced by Harris.

In 1887, Robert Craig assisted in "disseminating the productions of this eminent grower." Craig distributed 'L. Canning', 'Beauty of Kingsessing', 'Elkshorn', 'Mrs. G. W. Coleman', 'Mrs. A. Blanc', and 'Mrs. Wm. Howell', and Waterer sent out 'Wm. Dewar', 'Public Ledger', 'Stars and Stripes', 'Magnet', 'Mont Blanc', 'Colossal', 'Mrs. Sam Houston', and 'Miss Anna Hartshorne'. A variety called 'Mrs. Joel J. Bailey', which won "the fifty dollar silver cup offered by the Pennsylvania Horticultural Society," was also distributed, followed by 'Sunnyside', 'Mrs. T. C. Price', 'Mrs. M. J. Thomas', 'Mrs. John N. May', 'W. W. Coles', 'Mrs. A. C. Burpee', and others in the next year by Craig; Waterer distributed 'Excellent' and 'Robt. Craig', "also from the hands of Mr. Harris." In 1889, Waterer distributed 'Mrs. W. K. Harris', which took first prize offered by the Pennsylvania Horticultural Society for best seedling.

In addition, there were also 'Violet Rose', 'Ivory', 'Mrs. Irving Clark', 'Advance', 'Mountain of Snow', 'Miss Mary Wheeler', and others too numerous to allow individual mention.

3

Penstemon

PENSTEMONS ARE NOT universal favorites like roses or carnations, and some might question why they are included in this volume. They hold an important place in American horticulture, if only because hummingbirds find them so attractive. As gardeners move away from exotic imports and back to native plants this handsome native will become ever more popular.

The genus *Penstemon* has been subject to the usual taxonomic and etymological vagaries we have come to expect of widely grown garden plants. To start with, no one is quite certain why Dr. John Mitchell coined the term *penstemon* to describe the flowers he found in colonial Virginia and North Carolina. John Mitchell (1711–1768) was a physician and scholar who was educated at Edinburgh. When he returned to Virginia, he practiced medicine but spent a lot of time combing the Virginia and North Carolina countryside for new plants.

The flowers he discovered have a modified tubular structure with four active stamens and a fifth organ appearing to be a stamen but not producing pollen. In technical terms this is known as a staminode. Mitchell published

Wild species penstemon.
Photo by jgolby, Shutterstock.com

the first description of the new plant in 1748. Carolus Linnaeus included *pentstemon* in *Species Plantarum* in 1753. By botanical convention this is the Year One for nomenclature. (Anything coming before that is usually not considered.)

Fortunately, Mitchell reprinted his article in 1769, once again using the word "*penstemon.*" Because of this, the authorities eventually agreed that *penstemon* was correct. Mitchell's plant was *Penstemon laevigatus,* and the sample Linnaeus described was probably the same species.

Linnaeus and Mitchell had very cordial epistolary relations, and Mitchell accepted the fact that his plant was a *Chelone*, the genus to which Linnaeus assigned the plants, designating the species as *pentstemon*. Because of ill health, Mitchell spent a large part of 1745 in London. While there, he assisted the staff of the fairly new Royal Botanic Gardens at Kew with their new greenhouse. In 1748 he was elected a fellow of the Royal Society.

Regarding the etymology of the plant's name, at first it seemed obvious that since there were five similar organs, the syllable "*pen*"- was a corruption of "*pente*," the Greek word meaning "five." Linnaeus himself took this route, and called the genus "*Pentstemon*." The highly influential Liberty Hyde Bailey, the great Cornell University botanist and scholar, continued this usage. A few taxonomists even used the term *pentastemon*.

In 1966, a botanist who was also a classical scholar, Dr. Lloyd Shinners, offered another solution. He proposed that a different Greek root was at play, the word *paene*, meaning "almost," as in *peninsula* or *penultimate*. The fifth organ in these flowers was "almost a stamen."

A few new species were discovered by the end of the eighteenth century: *Penstemon campanulatus* was found in 1791, for example, and *P. barbatus* in 1794. Then the pace picked up. For the next twenty or thirty years, all the considerable number of new plants being discovered were put into the genus *Chelone*. Many of them were found in Mexico. Both North America and Europe were in the grip of major political upheavals during that period, and plant exploration was not a priority for those nations. Only by about 1820 did the idea that *Penstemon* was a genus unto itself emerge, and the genus *Penstemon* was created in 1828. John Fraser offered penstemon seed in his London shop in 1813, thought to be the very first time that was done.

More species were found in the eastern United States and the Plains region, but David Douglas, the Scottish plant explorer, found eighteen new species in the Far West. About 100 species from the Rocky Mountain region and the trans-Mississippi West were named by the middle of the nineteenth century. Some other species were discovered by John Torrey, a botanist traveling with the U.S.-Mexican Boundary Survey. Torrey was an important figure in the history of American botany. (Another botanist of immense stature, Asa Gray, was also attached to the survey team, although he did not find any new penstemons.)

Hybridization began quite early, but here again the history is somewhat murky. From what is known, crossing began earlier in London and other European cities than in the United States. The plant known as *Penstemon hybridum* was listed for sale by the firm of Flanagan and Nutting in London in 1835. The parents were not recorded, however, and it is not clear what this plant really was. Flanagan and Nutting also offered seed from nine other penstemon species.

Penstemon 'Pershore Pink Necklace'.
Reproduced by permission of Professor Dale T. Lindgren

The first historians to record these crossings suggested that *Penstemon hartwegii* and *P. cobaea* were the initial parents. In this, as in so much other penstemon history, a gentle skepticism is required. David Way and Peter James looked into the dates at which seed of these species became available in London in relation to the dates the new crosses were released. They showed that for the crosses to appear in the way that historians suggested would have been impossible. Among other reasons is the fact that *Penstemon hartwegii* seed was not sold in London until 1836, a year after *Penstemon hybridum* was listed for sale.

William Drummond, the brother of James Drummond (who found many significant plants in western Australia), first sent *P. cobaea* seed from Texas in 1836. Another distinguished plant collector, the Frenchman Henri Galleotti (1814–1858), had not even reached Mexico, where he would find *P. hartwegii,* by the time its seeds were said to be available. Sir Joseph Paxton, the Duke of Devonshire's gardener and creator of the Crystal Palace, described a new wild plant, *P. gentianoides* var. *splendens,* in the *Gardener's Chronicle* in 1842. This plant was identical to *P. cobaea.*

A Victorian garden writer, William Thompson, reported in 1851 that a French breeder, Alfred Pellier, had introduced a new hybrid, *Penstemon hartwegii* × *P. gentianoides*. Much more work was being done in France than in England. In Nancy, Victor Lemoine worked with penstemon (among his many other endeavors). He exhibited some of his penstemon hybrids at the Royal Horticultural Society's trials in 1861. More hybrids were introduced, and in 1870, *Floral Magazine* congratulated the firm of Downie, Laird, and Laing in Edinburgh for their wide range. These series of plants have come down to us as the European Hybrids.

Most of these cultivars have large red or purple flowers, making for a very colorful display in a border. Reginald Farrer, the eccentric English plant explorer and author known for his florid writing style, called them his "gorgeous garden fatties." One thing is clear. The European hybrids were, with one exception, derived from Mexican species, and not U.S. ones. Crossing the myriad of species from the western regions of the United States did not occur for almost another hundred years.

Penstemon 'Flathead Lake'.
Reproduced by permission of Dale T. Lindgren

BOTANY

Penstemon is a very large genus with many species. One national database, the Germplasm Resources Information Network (GRIN) of the U.S. Department of Agriculture, lists 240. Another estimate is 270 species. The majority are found in the Western Hemisphere, particularly in the western United States. Botanists recently changed the family to which penstemon belong. After many years in the Scrophulariaceae, penstemon is now in the Plantaginaceae. These decisions are based on recent work matching DNA patterns, the new science of cladistics.

The genus is divided into subgenera, sections, and classes. The subgenera include *Dasanthera, Saccanthera, Cryptostemon, Dissecti, Penstemon,* and *Habroanthus. Dasanthera* species, distinguished by their wooly anthers, are most commonly used in breeding commercially successful hybrids. The *Penstemon* subgenus has 186 species, whereas *Dasanthera* only has 9.

Penstemons crossbreed very easily in the wild, and there are many naturally occurring hybrids. In 1930, Anna Johnson in Butte, Montana, introduced a very attractive coral red penstemon. A collector told her he had found it growing close to the shores of Flathead Lake. She subsequently released it for sale with the name "Flathead Lake Hybrid." It became very popular and is still available. The plant has been used as the basis of several other intentional hybrid series, as I discuss below.

The original guesses at the parentage of this plant were not borne out by later study, but in the 1960s, it was called *Penstemon × johnsoniae* in honor of Anna Johnson. *P. barbatus* is one parent. It seems probable that the initial plant was the result of a chance cross in a Montana nursery, yielding a hybrid that escaped and naturalized successfully.

BREEDERS

UNITED KINGDOM

John Forbes (1842–1909), Hawick, Scotland

Forbes was a nurseryman in Hawick, a relatively small Scottish border town. He started his first venture by renting a fairly small place in the district of Dovemount in Hawick, in 1870. In 1878, he rented five acres between the brewery and the mill and expanded his premises. His firm, called Buccleuch Nurseries, opened the following year, and its first catalogue appeared in 1880. The land belonged to His Grace the Duke of Buccleuch, for whom the nursery was named.

John Forbes's father, Archibald, had been a gardener in the village of Logierait, Perthshire, and a bagpiper too. Piping for the gentry as well as tourists augmented Archibald's income very satisfactorily. John was born in a cottage on the estate at Killiechassie House in Aberfeldy, a market town on the River Tay in Perth and Kinross, and was apprenticed to Mr. Balfour, the head gardener there. John clearly was an apt student and willing to work very hard. After further training he reached Hawick in 1865 and became head gardener to David Pringle at Wilton Lodge. Mr. Pringle treated him generously, allowing him to grow plants for his own use in the lodge's walled garden. The garden is now open to the public, and the land around Wilton Lodge is a park.

Forbes was ambitious. Within five years of starting to work for Mr. Pringle, he opened a small nursery. He rapidly prospered and later developed extensive premises, with twelve greenhouses. When or why he became interested in penstemons is not known, but eventually his nursery introduced more than 550 cultivars.

He even exceeded Victor Lemoine in this sphere. They seem to have had a friendly rivalry and Lemoine allowed Forbes to use many of his plants as a basis for new crosses. Lowland Scotland is not really very much like the places where penstemons are found, but its climate was mild enough to allow the plants to flourish.

John Forbes was president of the Hawick Horticultural Society at one point and a town councilman for a few years. Forbes and his wife had six children, one of whom died in infancy. Two of his daughters worked in the office at the nursery, and his son took the business over after his death. The firm lasted until 1969. In 1907, two years before he John Forbes died, he was granted the King's Warrant to supply seeds and plants to Balmoral, Windsor, and Sandringham, as well as the royal parks. At his peak, Forbes employed twenty-eight men and boys for the gigantic tasks of propagation and bedding out. The author of an admiring article noted that the firm bought more than three tons of small flowerpots every year.

Forbes did not hesitate to charge very high prices for his special plants, and the pansies in particular brought in enough money to cover his payroll quite comfortably. The Buccleuch Nurseries won medals at the St. Louis Exposition in 1904 for their penstemons as well as other flowers. Their 1908 catalogue announced twenty-five "New Giant Penstemons for 1908." Forbes introduced "Grand New Bedding Penstemons" such as 'Newbury Gem', 'Crimson Gem', and 'Newbury White Gem'.

Some of Forbes's penstemons are still known. The medium-sized, half-hardy ones used for bedding have survived and are available. His other significant contribution was the very tall 'Florist Giant Penstemon'. These are wonderful plants for a border, as well as being useful as cut flowers. This cultivar is no longer extant, but George Thorburn, a former employee of Forbes and a horticultural expert, believes he could reproduce it through crossbreeding with F1 techniques.

Thomas Brown, a unique specialist in the reconstruction of historical landscapes, collected information about plants introduced before 1900. He only lists three of Forbes's penstemons, one from 1888, the others from 1897. Dale Lindgren's larger and more comprehensive list, covering not only the nineteenth century but also the twentieth, shows about a dozen of them. These were introduced between 1901 and the 1930s, after John Forbes had died. He had worked closely with his manager/foreman, William Oliver. Both were expert plantsmen.

Hawick Museum owns a Buccleuch Nurseries catalogue from 1910. Under penstemons, it notes that no plant can compete with the autumn display of Forbes Extra Choice Mixed Penstemon: "The seed offered has been saved from our world renowned collection, which cannot be excelled."

Looking back, it is amazing that an enterprise of this scale and quality should have flourished so well and then been almost totally forgotten by the larger world. The business fell prey to the now-classic shifts in the global horticultural trade: the vastly improved and rapid transport of fragile things like flowers, the movement of jobs to countries where wages were a fraction of those in Scotland, and the establishment of nurseries in warm countries to avoid the costs of heating greenhouses.

FRANCE

Victor Lemoine (1823–1911), Nancy

Lemoine was a giant figure in the nineteenth-century flower breeding world. Very few other people even came close to the number of introductions he made across a broad spectrum of ornamental plants, both herbaceous and shrubby. In the United States, he is remembered fondly for his "French lilacs," inspired crosses initially between *Syringa vulgaris* L. 'Azurea Plena' and a cultivar of *S. oblata*. He started doing this in 1870 to take his mind off the ghastly events of the Franco-Prussian War. Paris was under siege, and many people starved. Everyone in France suffered abominably, including the Lemoines in Nancy. Victor Lemoine was the son and grandson of estate gardeners in Delmes, Lorraine. The family could afford to let him stay in school for the full course. After that, he spent a year working at three different nurseries in eastern France and Belgium, learning as much as he could. His father lent him money to buy a small plot of land in Nancy, a rapidly growing silk manufacturing city, and he opened his nursery in 1849. By 1854 he had introduced his first hybrid, a double purslane.

Lemoine became famous quite young, and received many local and international honors. In 1894 the Royal Horticultural Society in London gave him the Veitch Medal, which has only ever been awarded to a handful of foreign horticulturists. A very enthusiastic American amateur gardener, Mrs. Bechhold, actually named her son Lemoine Bechhold in his honor. Young Lemoine Bechhold grew up to hybridize *Hemerocallis* as an avocation.

The bright and assertive modern "Zonale" pelargoniums came from Lemoine's nursery, as did countless begonias, gladioli, peonies, and chrysanthemums. Most modern gardeners are not aware that Lemoine and his descendants brought out more than four hundred penstemons. Very few of them, if any, are still in commerce. Lemoine used the same primarily Mexican species for his crosses as did the other European nurserymen. He issued the first cultivars in 1860. His son Emile, together with his widowed mother Mme Lemoine, continued the work after Victor Lemoine died in 1911. The last cultivar, which came out in 1934, was bred by Lemoine's grandson Henri.

Lemoine corresponded with John Forbes about penstemons and often sent him breeding material for the latter's experiments. Unfortunately, the Musée Lemoine in Nancy has no pictures of the flowers or records of the correspondence.

Alfred Pellier

Pellier was a French breeder who introduced a new hybrid penstemon in 1855. In 1860 Victor Lemoine named a penstemon for him.

GERMANY

Wilhelm Pfitzer (1821–1905), Fellbach, near Stuttgart

Pfitzer was the son of a harness maker, and his parents allowed him to use some of their land near Stuttgart. As a boy, Wilhelm had shown an eager interest in nature, and his parents encouraged him. In 1844 he decided to open a nursery and devote himself to horticulture. Pfitzer was attracted to penstemon as early as the 1850s, and by 1857, he was offering twenty-four varieties.

In the hands of Pfitzer's children and grandchildren, the firm was very successful for more than a century, but by the 1980s it had shrunk materially. Two world wars and the huge upheaval in post–World War II Germany were a large part of the problem, but the shift of the floral industry to warmer climates with their cheaper wages was the finishing touch. (See chapter 4 for more about the Pfitzer family.)

Bronze memorial medallion of Wilhelm Pfitzer in Germany.
Reproduced by permission of Bernd Kaiser

Wilhelm Pfitzer, founder of a family nursery firm in 1844 that continues to this day.
Reproduced by permission of Klaus Pfitzer

Wilhelm Pfitzer II (1854–1921)

Pfitzer's son was also an accomplished plant breeder. His grandson continues the family tradition but as an expert in dahlias.

UNITED STATES

Glenn Viehmeyer (1900–1974), Nebraska

Viehmeyer was born in Logan County, Nebraska, and lived in the state all his life. He worked at the University of Nebraska's North Platte Experiment Station from 1943 to 1966, ultimately introducing fifty new varieties of flowers, fruit trees, and shrubs. Some of his other contributions were the use of phosphorus fertilizers and chemical controls for pests. Viehmeyer really began modern penstemon breeding at the University of Nebraska at Lincoln.

Glenn Viehmeyer. *Reproduced by permission of University of Nebraska–Lincoln*

Penstemon 'Prairie Splendor'.
Reproduced by permission of Professor Dale T. Lindgren

In 2010, Dale Lindgren, historian for the American Penstemon Society (APS) observed,

> When Glenn began his efforts to cross wild species of penstemons, it was the firm belief of our members that penstemons were one genus of wild flowers which it was impossible to cross. Up to that time the only successful crosses had been one between *P. grandiflorus* and *P. murranyanus*. It was almost taken for granted that no other species would cross, but Glenn, ignoring this traditional belief in the society, started experimenting and soon discovered a remarkable fact, that one form of penstemon that we had been growing for many years, under the name of 'Flathead Lake', pos-

sessed the almost unbelievable capability of accepting pollen from almost any other species. The resulting crosses, moreover, also proved capable of being crossed with other species. The result of this remarkable discovery was that Mr. Viehmeyer began coming out every year with new crosses between species which we have never considered capable of being crossed, thus putting an end to the notion that penstemon species will not cross.

One of Viehmeyer's earliest efforts was crossing the 'Flathead Lake' hybrid with *Penstemon strictus*, to produce *P.* 'Prairie Dusk'. He followed it with *P.* 'Prairie Fire', created with pollen from one of Lena Seeba's Nebraska hybrids. A third one was *P.* 'Prairie Dawn', using pollen from a complicated hybrid: (*P. glaber* var. *alpinus*) × (*P. clutei* × *P. palmeri*) and *P. kunthii*.

Viehmeyer's hybrids remained in commerce for a long time. Three other enthusiasts introduced hybrids that were very attractive at the time but rapidly faded from the market. Lena Seeba was one of them. In the late 1940s, she found a plum-colored sport of *Penstemon grandiflorus* × *P. murrayanus* and used it as the basis of her work. Fred Fate issued many hybrids, and Alan Scharf of Saskatoon crossed 'Prairie Fire' with varieties of *P. cardinalis*.

Gardeners have every reason to be grateful to men like Glenn Viehmeyer even if they do not know it.

Gladiolus

IT IS HARDLY surprising that a tall handsome plant with strong tapering leaves should be named *Gladiolus*, diminutive of the Latin *gladius*, "sword." Linnaeus coined this name in *Species Plantarum* (1753). Gladiolus has a fairly wide distribution in Mediterranean Europe, Asia, and tropical Africa, but the center of diversity is the Cape Province in South Africa. Of the approximately 260 species, 10 are European and Asian. All the rest are African.

The nomenclature has undergone numerous revisions. Readers are advised to take a deep breath as they go through the next section. The beautiful modern hybrids, which we can buy so easily nowadays, have many antecedents, and yet really only about a dozen of the available species have been intensively crossed. In the past few years, breeders have begun looking into intergeneric crosses to breed in new desirable traits.

One European species, *Gladiolus communis* Mill., was recorded as early as 1575. John Gerard mentioned it in his *Herbal* in 1597. Another species, *G. byzantinus*, was mentioned by John Parkinson in 1629. It was called the Corn Flag of Constantinople in those days, then received a new name and is now called *G. communis* subsp. *byzantinus* (Mill.).

Gladiolus primulinus, now *G. dalenii* Van Geel.
Photo by lokvi, Shutterstock.com

By the mid-eighteenth century, the African species *Gladiolus blandus* Aiton and *G. tristis* L. were available in Europe. The current name for *G. blandus* is now *G. carneus* D. Delaroche. *G. tristis* was found in Natal, in South Africa. English and Dutch merchant ships plying the India trade carried these two species, as well as *G. recurvus* L. and *G. alatus* L., between 1739 and 1745. Almost all the ships stopped at the small port of Cape Town in South Africa to take on supplies for the long journey back to northern Europe. Philip Miller, head gardener at the Chelsea Physic Garden and author of the first true comprehensive encyclopedia of gardening (c. 1731), gave *G. tristis* its name.

Gladiolus cardinalis and another form of *G. blandus* reached Europe in 1789. William Curtis took note of it in his *Botanical Magazine,* and indeed *G.*

cardinalis is now known as *G. cardinalis* Curtis. Richard Anthony Salisbury presented three papers about it at the Horticultural Society in 1812. The flower is very striking, with rich crimson and white petals. The wild plant grows beneath waterfalls and thus is very hard to maintain in an ordinary garden. Another South African species, *Gladiolus psittacina* (now *G. dalenii* Van Geel but with another previous synonym, *G. natalensis*), was noted in the *Floricultural Cabinet* in 1834.

The plant became popular in Europe after Napoleon III's gardener at Fontainebleau, Eugène Souchet, showcased it in his displays. Even when the only specimens were still rather thick and stiff, the fact that the flowers were well formed and open at the same time along the stem was very attractive. They also came in delightful colors. Queen Victoria visited Fontainebleau in 1853 with Prince Albert and was enchanted by the flowers.

Souchet had cunningly distributed vases of them among his borders to draw the visitors' eye. The queen succumbed, and requested that some of these plants be sent to her gardeners at Osborne. Once she went home to London and everyone heard about her enthusiasm, English nurserymen got busy vying for the royal patronage.

A species with a pale yellow flower, *Gladiolus primulinus* Baker, was discovered in 1903 near the Zambezi River in Africa. Stephen F. Townsend, an engineer working on the construction of the Victoria Falls Bridge, sent corms to Francis Fox in London, who exhibited specimens the next year. *G. primulinus* is now one of the synonyms of *G. dalenii* Van Geel. Species gladioli flower at different times throughout the year, but the primulinus type are summer flowering.

HYBRIDIZING

Although much more was happening on the Continent in the early nineteenth century, a small amount of hybridizing did take place in England. One early enthusiast in the United Kingdom was Dean Herbert (see below).

The first commercially viable cultivars were introduced by the firm of James Colvill (1746–1822) of Hammersmith in the 1820s. Colvill opened his nursery in 1783 and took a Mr. Buchanan as partner. They built almost 40,000 square feet of greenhouses. Colvill's son, also named James, took over when his father died. Colvill's hybrids were known as *Gladiolus colvilli*, and *Gg.* 'Roseus', 'Rubra', and 'Albus' are said to be available still. A very popular Colvill hybrid was 'The Bride', a dwarf white form. These are hybrids of *G. tristis* var. *concolor* and *G. cardinalis*.

Hermann Josef Bedinghaus (1806–1884), Enghien, Belgium

Hermann Bedinghaus, the duc d'Arenberg's head gardener at Enghien, near Brussels, introduced the Ghent hybrid *G. gandavensis* in 1837. His work was taken up very quickly, and new hybrids began to appear as early as 1841. *G. gandavensis*'s parentage was said to be *G. oppositiflorus* and *G. natalensis* (now *G. dalenii*). The correct name for this hybrid is *Gladiolus* × *gandavensis* Van Houtte. The term *gandavensis* signifies the plant's origin in Gand (Ghent), reflecting Van Houtte's influence.

Bedinghaus did not remain the Duke of Arenberg's gardener all his life. He built up a highly esteemed nursery in Nimy, near Mons, quite probably after he had saved money and formed useful connections. He became very active in the Société Royale d'Horticulture de Mons and the Société Horticole et Agricole du Hainaut, the province where he worked. There was a very fine tribute to him in the society bulletin in 1878: "Un de nos plus anciens et émérites horticulteurs, dont le nom est bien connu depuis longtemps" (One of our most distinguished and best-known horticulturists for many years [my translation]).

In 1863 the German botanist Karl Heinrich Koch named a Mexican plant for him: *Furcraea bedinghausii* (now *F. parmentieri*). Bedinghaus had been growing a specimen of this plant in his greenhouse and knew that other growers used various different names for the same plant. It looks like *Agave* but has enough taxonomic differences to put it in a different genus, though it is still in the family Agavaceae. Bedinghaus entered his plant into a show and asked Koch to confirm his impression that it was indeed *Furcraea*.

A massive member of the genus, *F. longaeva*, grows at high altitude in Mexico, lives for a very long time, and has an enormous florescence. In its maturity it can be compared to a feather duster on a very long trunk. The reason to go into this much detail is to demonstrate Bedinghaus's seriousness and conscientious respect for science. After releasing *G.* × *gandavensis,* he moved on to other work without apparently looking back.

Gladiolus × *gandavensis*

Almost immediately, Louis Van Houtte bought the exclusive rights to Bedinghaus's hybrid, which had a much larger flower than its nearest relative, *Gladiolus ramosus,* and snapped up all the seedlings. Van Houtte had built the largest nurseries in Europe and was a very successful businessman at a time when Belgium was an international force in horticulture. *G. ramosus* is no longer considered to be in the gladiolus genus and is now *Melasphaerula ramosa* (L.) N. E. Br.

At the time, feelings ran high because Van Houtte trumpeted this variety as his own work, giving no credit to anyone else. Van Houtte also wrote, printed, and published the magazine *Flore des serres* for many years. Much of its content was openly plagiarized from other publications, yet there do not seem to have been many complaints. In one egregious instance, he somehow managed to get hold of a new English piece and brought it out before it even appeared in its official form. Readers were indulgent because he employed master engravers and filled the pages with elegant illustrations.

One man who did not forgive Van Houtte was the very upright, Very Reverend Dean Herbert in Manchester. He was offended by the latter's claim to the Bedinghaus hybrid. William Herbert, dean of Manchester (1778–1847), was a prominent member of the English horticultural fraternity and an accomplished plant breeder when it first became acceptable to do this. He was particularly interested in bulbous plants.

Herbert crossed the same species as Bedinghaus, but his hybrids did not do well: none of them set seed. In spite of that they are of considerable academic interest, because Herbert kept meticulous records of both seed and pollen parents, enabling each plant's lineage to be traced accurately, something that is not always the case. The Belgian variety was very important because it could be used to breed many more cultivars. It is truly one of the founders of the modern gladiolus, *G. dalenii. G. dalenii* is tetraploid, whereas most gladioli are diploid. Its descendants are also tetraploid and thus have the desirable large flowers.

Other *Gladiolus* species

Herbert commented that some of his seedlings were taken to Australia by the English plantsman John Bidwill (1815–1853). They thrived there and gave rise to new Australian forms. Bidwill is remembered primarily for his plant exploration, but he had learned to cross plants from his master, a man named Pince, in Cornwall as a very young man. Alas, he is also remembered for his complete insensitivity to the ways of the native Australians, which led to his being killed in the bush. Other species that have contributed to the modern flower are *Gladiolus aurantiacus, blandus, byzantinus, cardinalis, cruentus, dracocephalus, primulinus, psittacinus, trimaculatus, tristis,* and *tristis* var. *concolor*.

    G. aurantiacus, dracocephalus, and *trimaculatus* are obsolete names. Wilhelm Klatt called it *G. aurantiacus* in 1867, and the name is still in use in some quarters. *G. dracocephalus* was named by Joseph Hooker in 1871. *G. trimaculatus* goes back to 1788, when Lamarck named this plant; it is a synonym of *G. carneus* F. Delaroche. The species now known as *Gladiolus dalenii* has seven former synonyms, a rather bewildering number: *cooperi, leichtlinii, natalensis, primulinus, psittacinus, psittacinus* var. *cooperi,* and *quartinianus*.

    You may now exhale. The gladiolus turned out to be very adaptable and is grown around the world. For example, it flourishes in India, and numerous cultivars suited to the Indian climate have been developed at the Indian Institute of Horticultural Research. Australians and New Zealanders grow gladiolus enthusiastically, and there are gladiolus societies in both those countries.

    The names that follow were partially derived from the comprehensive records kept at Cornell University, first by Professor Alvin Beal and later by his successor, Professor Alfred Hottes. Cornell University is a private foundation, but the college of agriculture was a land grant college, and the legislature of New York State required an annual report from the agricultural extension about new crops and work with existing ones. Several floral organizations, among them the American Peony Society and the North American Gladiolus Council, took advantage of the university's open land and skillful staff to do their testing and trials.

    The first two volumes of *Gladiolus Studies* appeared in 1916, prepared by Alvin C. Beal, professor of horticulture. After Beal died, his successor, Alfred Hottes, issued the third volume. Both men were consummate scholars, and

Alvin Beal.
Reproduced by permission of Cornell University

we are in their debt for a job very well done with no obvious reward in sight. They probably could not imagine that a hundred years after their death, their work would still be relevant.

Beal kept thorough and meticulous records of each grower and the flowers they entered, together with a thorough history of the gladiolus up to 1915, while Hottes listed all the known cultivars and their creators (including all the cultivars in the trials). He recorded the dates at which the cultivars were introduced.

I turned this information into a database and sorted it by the number of cultivars introduced by each grower. As noted in the previous chapters, I made a slightly arbitrary decision that anyone who left eight or more new cultivars was a serious grower. That was the threshold for including a grower in this list.

Extracting the data from Hottes's lists provided a very clear window into gladiolus growing at the time. When these names were added to those in lists compiled by the late Thomas Brown of Petaluma, California, titled "Gladioli b4 1900," almost 150 growers emerged. Men like Arthur Cowee and Henry Harris Groff appeared on both Brown's and Hottes's lists.

No review of the North American gladiolus scene a century ago would be complete without mentioning Madison Cooper of Calcium, New York. The town is fittingly named, as he started in the dairy business. Despite his Anglo-sounding name, Cooper was a direct descendant of a nineteenth-century French pioneer, Pierre Coupier.

Madison Cooper became widely known for his skill in refrigeration, and worked in the wholesale dairy business for many years. Then, quite abruptly, he felt he was wasting his life and wanted to get back to his roots in the small upstate New York town where he grew up. He bought land and began growing gladioli. To make ends meet, he had the happy idea of starting a publication, *The Modern Gladiolus Grower,* in which growers could explain what they were doing, and act as a forum for anyone who was interested.

The magazine went through numerous name changes but survived until the 1930s, when Cooper had to sell it. Even after that, he was still drawn to the publishing business, and started yet another version of his magazine, but it lasted only a few years. The original version of the magazine is a magnificent resource for anyone wishing to find out what was happening in the gladiolus world in the early twentieth century. Cooper found contributors from all over the world, including such remote places as the island of Guernsey in the English Channel.

CANADA

Henry Harris Groff (1853–1933), Simcoe, Ontario

Groff was born in Simcoe, a smallish town in the southwestern part of Ontario, and lived there almost his entire life. The son of the local banker, he first studied pharmacy with the town druggist but later joined the Federal Bank and worked his way up into management. After the bank closed, he developed a private banking business.

Groff became intrigued by the breeding of flowers as a very young man after dabbling in breeding birds and small mammals. (There are some similarities to W. Atlee Burpee's initial career in in poultry and poultry feed in this story.) Groff crossed his first gladiolus in 1880. He went on quietly cross-

Henry Harris Groff.
Source: Who's Who in Canada *(1914)*

ing more and more gladioli, until he decided to enter the Pan-American Exposition at Buffalo, New York, in 1901. The result was utterly unexpected. Everyone was taken completely off guard.

Despite being totally unknown in the insular world of gladiolus specialists, Groff won a clean sweep of the prizes at the Pan-American Exposition show. Luther Burbank congratulated him, in a rare moment of humility and recognition. Burbank was used to receiving adulation, not showing it. Over the next twelve years, Groff swept all the awards for gladioli in the United States, Canada, and many other countries.

Groff was not working in a vacuum. He was fully immersed in all the available science and laid his foundation very carefully. Groff traveled widely in pursuit of many varieties with which to experiment, and few experts knew as much as he ultimately did. He bought Luther Burbank's entire stock and later that of Walter Van Fleet, a distinguished horticulturist in New Jersey. Van Fleet later went on to horticultural immortality by breeding exceptional roses, but at that time he pursued the gladiolus energetically.

A senior official at the U.S. Department of Agriculture considered Groff to be a leader in the field of plant breeding. Groff surprised everyone when he said, contrary to current opinion at the time, that he was able to derive the vigor and vitality of his new cultivars from hybrid sources and did not have to go back to wild germplasm. Journalists of the time called this a "bombshell."

Groff met Arthur Cowee of Berlin, New York, at the Buffalo show and retained Cowee as his North American agent to distribute his plants. Cowee was a coal merchant with a penchant for gladioli. His hobby was gardening and raising flowers, but later the hobby changed to being a primary source of income. Cowee knew about Groff's work because he had grown some of the latter's gladiolus seeds satisfactorily.

Through Cowee and his business connections, orders came in for Groff corms from all over the world, and Groff is said to have shipped two million cormels to Cowee over the following two years. What happened next once again took everyone by surprise, just as Groff's sweep of the awards at the Pan-American Exposition had done.

A few years after his astounding success and having made a great deal of money from it, he gave up breeding gladioli completely and returned to the banking business. His gladiolus career peaked in 1914. In horticultural terms, we can say he became dormant. He felt he had gone as far as he could with the gladiolus and preferred to retire from the field rather than cast about for new challenges to meet.

In the mid-1920s he just as suddenly came back to life, again horticulturally speaking. He fixed upon the iris as a suitable vehicle for his restored energies and went on to win new prizes and championships with that flower. Groff married Ellen Skynner of Toronto in 1879, but there is no record of any children. His obituary only mentions a sister and a niece and nephew as his survivors. There were no descendants to consolidate his accomplishments and thus no way to find out how he reached his seemingly extraordinary decisions. Henry Groff was truly a unique person and it would be wonderful to know more about him.

ENGLAND

James Kelway (1815–1899), Somerset

Kelway was a very prominent English nurseryman in Somerset. His business is still extant today, though very much modified. He is perhaps best known for his delphiniums, but gladioli interested him right from the start of his career. He was only fifteen when he entered a group of plants at the local flower show in Glastonbury and won most of the prizes. Kelway first saw a gladiolus at that competition, and had a similar response to that exhibited by many other gardeners upon being introduced to the plant. A suitable word for such a response might be joy.

The Duke of Devonshire is perhaps the best known in this set of mainly men and boys. They latch onto a plant as though an image of it had been waiting in their heads all along. Seeing the actual one triggered the reaction.

The duke's obsession was with orchids. A life that had seemed meaningless suddenly blossomed with purpose. The French have an expression for this: "un coup de foudre" (as if hit by a thunderbolt). (See *Visions of Loveliness* for more about the duke.)

This reaction is not all that different from a gift for music emerging when a young child hears a particular instrument for the first time. The case of Jacqueline du Pré and the violoncello is well known. She heard it being played when she was four and gave her parents no rest until they bought her a cello. She then rapidly proceeded to learn how to play it and received great acclaim.

The son of a head gardener at Westholme in Somerset, James Kelway was also a head gardener when he decided to go into business for himself. This was often the way successful nurseries began. In 1850, at the age of thirty-five, he bought land in the parish of Huish Episcopi. In spite of the slight geographical inconvenience, he always called his firm Kelway's of Langport instead. Langport was actually the neighboring village. One can see his point: Huish Episcopi is a bit of a mouthful, even for the Church of England faithful in Trollope land.

Eugène Souchet was thought to be the greatest breeder of gladioli alive at the time and had won the first prize for his gladioli at the Paris International Exhibition in 1867. This provided Kelway with inspiration and an exciting challenge. Kelway met Souchet in 1874, and they became friends. Kelway crossed some Lemoine hybrids with the Gandavensis hybrids and came up with the "Kelwayi" cultivars.

The Kelway firm entered many of their unique Kelwayi cultivars in the trials at Cornell University in the years leading up to 1914, at the same time as their American counterparts. Alfred Hottes, Alvin Beal's successor, documented these and all the other entrants in the trials in volume 2 of *Gladiolus Studies*.

The Kelways also participated in the Panama Pacific International Exposition in San Francisco in 1915. They were one of a handful of British nurseries that submitted entries and they won a medal of honor. The United Kingdom did not have a national pavilion at that exposition, unlike Japan, France, the Netherlands, and Turkey. Kelway's entries were individual and independent. The original notes made by flower judges at the exposition have been preserved, and one can see how they made their decisions.

Someone at Kelway's used *G. primulinus* early to create the Langprim series. The flowers were triangular and hooded on a tall slender stem, and the color range was extended yet further into yellow, orange, salmon, cream, and ivory.

In 1873, Kelway's had twenty-five acres in gladiolus, including some eight hundred cultivars. By then, James had taken his only son, William, as a partner. Because of William's unpleasant behavior the nursery staff heartily disliked him. He was said to be too strict and to spy on them unseen. William's son James inherited the nursery after the disasters of the First World War. Trying to make a living from a nursery became almost impossible and James finally had to admit defeat.

He was declared bankrupt in 1934 but was retained as manager by the consortium that later bought the business. He died suddenly in 1952. The business then went through more than one fresh start as three new owners tried to make it work. The final owner, Dave Root, arrived in 1995, and the business has remained stable since then under his steady guidance. No Kelway family members participate in the present firm.

The rise and fall and ultimate re-creation of Kelway's illustrates again the surefootedness with which a capable man could build a successful business and go from strength to strength in a stable and unchanging world. No one had to tell James Kelway how to proceed. The necessary steps all seemed to unfold in his head. The tragedy was that men such as this were not nimble in the face of change. Very few are.

Frank Unwin (1898–1988), Histon

Frank Unwin, of Histon, near Cambridge, was the younger son of W. J. Unwin of sweet pea fame. (See *Visions of Loveliness*.) W. J. wanted his son to work in his business, but Frank abhorred being stuck indoors. When given the choice of digging a four-acre lot for an orchard by hand rather than sitting in an office, he chose the former backbreaking task. Although Englishmen are notoriously undemonstrative, his father loved him deeply and bought him his own farm a few months before the First World War erupted.

Frank Unwin was another man who was smitten by the gladiolus when very young. In 1923, he met a Dutch breeder, Klaus Velthuys, on a visit to the Netherlands with his father. Velthuys was working with *Gladiolus primulinus*. Nothing for it but Frank had to start breeding his own varieties with this exquisite flower too, even though farmwork left him almost no spare time.

Frank worked with *G. primulinus* soon after it became available, using it both as a seed and a pollen parent, whereas Kelway had used it solely as a pollen parent. These more delicate flowers were more popular than Kelway's and remained in cultivation for much longer.

Frank Unwin moved back to be with his father in Histon after Frank's wife died very young, and they worked together for many years. Unwin Junior was president of the Gladiolus Breeders' Association for a long time. A woman who worked in the firm many years later remembered him as a kindly elderly gentleman who always brought flowers for the staff.

Channel Islands—Jersey and Guernsey

These islands between the coasts of England and France have a reliable and equable climate because of their proximity to the Gulf Stream. Many flowers and vegetables have been grown there successfully, and gladioli were no exception. In the 1850s, nurserymen bred some of the first Nanus cultivars, probably using *Gladiolus cardinalis* and *G. venustus* as parents. Colvill's 'The Bride' was a white form of this dwarf class. There is very little literature on this topic.

In a fortunate referral, I was introduced to a more recent family of gladiolus breeders in Guernsey, the Mahys. John Mahy's father added gladioli to his farm in the early 1930s, using a wild variety growing nearby as one of his breeding parents. (The use of the term "wild" is misleading, because gladiolus is not native to the Channel Islands. Sometimes seed of a cultivated plant establishes itself in the wild.) One of the other parents was 'The Bride'. From this crossing, Mahy Senior developed the cultivar 'Pure Bride', a flower with pure white stamens and pistils. He also introduced 'Amanda Mahy', 'Rita Page', 'Guernsey Glory', 'Blushing Bride', 'Fusilier', and 'Miranda'.

When the Germans invaded the Channel Islands in 1940, they requisitioned all available land and parked their tanks in Mahy's flower fields. After the Germans left, no traces of the gladioli reappeared. There was complete devastation. The Nazis' brutal behavior in the Channel Islands is not as well known as it should be.

What followed was a wonderful tale of reciprocity and simple generosity, gratefully remembered by Mahy Senior.

Before the war, a Dutch salesman, Simon De Goede from Elst, near Arnhem, had recommended taking the gladioli corms to the Netherlands to bulk up, as they could thrive in the lighter soil there. Mahy wanted to increase his stock, and accepted De Goede's arrangements. For the next few years, half of the Mahy stock went to the Netherlands.

After the war, Mr. De Goede once again called upon the Mahy family, and they told him about the destruction of their corms by the Germans. He immediately promised that he would give them half of everything, which had done very well in the Netherlands, if he could help get the Mahy business going again. (One of the things that my correspondent did not marvel at was the fact that both Mahy and De Goede survived the terrible war.) The Mahys used the restored gladioli corms to grow flowers for the cut flower market, but eventually the corms grew weak and had to be discarded. By then the Mahys had switched to freesias as more economical to grow and thus more profitable.

FRANCE

Victor Lemoine (1823–1911), Nancy

Lemoine has been covered thoroughly in other sections of this and my previous book. He started with the Gandavensis hybrid but used *Gladiolus purpuroaureatus* and *G. dracocephalus* as the other parents. Unlike many of the Lemoine introductions, these cultivars were not particularly effective or long-lived. Contemporary reports expressed doubt about them at the time. Perhaps their principal worth lay in the wide range of colors that emerged. Not only were shades of blue, mauve, and violet available for the first time but also, when he used *G. dracocephalus* as a parent, brown and greens were possible.

Many years later, in 1946, the Dutch used the smoky Dracocephalus hybrids to revive their gladiolus industry. Nothing Lemoine did was ever wasted. His son Emile eventually took over Victor's work on gladioli and became a well-known expert on the flower.

Eugène Souchet (1812–1880), Paris

Many consider Souchet, Napoleon III's gardener, to be one of the greatest flower breeders of his epoch. He worked with *Gladiolus gandavensis* and introduced hundreds of new cultivars between 1850 and 1880, when he died. Souchet used *G. floribundus* and *G. ramosus* to create many of his cultivars. 'Ceres' and 'Eugene Scribe' were introduced in his lifetime, but quite a few of his cultivars were not released until after his death. The Vilmorin firm took over his stock and introduced new gladioli as Souchet-Vilmorin. This may have been in conjunction with Souchet's nephews, Messieurs Souillard and Brunelet, who continued his hybridizing work.

UNITED STATES

Tracing the course of gladiolus history in the United States has led me to many small towns, often in the Midwest, where a dedicated amateur has quietly worked wonders far away from the distractions of more exciting places. This is also a good time at which to say how much I appreciated the efforts of librarians, archivists, and genealogists in those towns, carefully treasuring the links with the past. The unexpected reconstruction of a formerly extensive business with national or even international reach that has completely disappeared is the reward for pursuing these figures of the past.

Gladioli arrived in North America as early as 1806, but did not become a popular garden plant for many more decades. Beal, at Cornell, noted that Bernard M'Mahon, a very early American nurseryman in Philadelphia, listed only European varieties at the end of his useful book in 1806. The very well known nurseryman William Prince offered fourteen species of gladiolus, including *G. tristis* (whose name means "sad gladiolus"), at his Linnean Gardens in Long Island in 1825. Most of the other nursery catalogues of the day did not yet carry them.

A more general interest in this flower began stirring in North America around the time of the Civil War. References to gladioli appeared sporadically in the horticultural literature of the time. Papers were read about individual experiences with particular species and from time to time a few gladioli were entered in flower shows.

The nurserymen persisted based on this public interest. David Landreth in Philadelphia and William Hovey in Boston were active, but Beal could not find any evidence of new cultivars being developed in North America at the time. Not until 1863 did the situation change, when an exciting group of new cultivars was exhibited at the Massachusetts Horticultural Society by E. S. Rand and others. Members competed for medals and prizes. This all led to the appearance of many new seedlings and a very thriving gladiolus scene.

Here is the conscientious Beal:

> E. S. Rand, Jr., as chairman of the floral committee of the Massachusetts Horticultural Society, published with his report for 1858 a paper on the culture of the gladiolus, in which he expressed the hope that seedlings would be raised. It appears later that Mr. Rand and others acted upon the suggestion, for the following statement is found in the history of the above-named society: "This year [1863] witnessed the commencement of those profuse and beautiful displays of seedling gladioli."

James McTear, a Massachusetts plant breeder from Roxbury who had joined the Massachusetts Horticultural Society in 1857, showed nine new cultivars on August 29, 1865, and one on September 12. George Craft, who worked in Brookline and joined the Massachusetts Horticultural Society in 1863, won the silver and bronze medals that same year. Beal noted that Craft's silver medal winner, 'Elnora', "was a pure white, in some cases faintly flaked with violet, the center petal feathered maroon on delicate lemon ground; it was characterized by a bold spike, a large flower, a neat and compact face, and vigorous habit."

Craft's 'Colonel Wilder Wright', which won the bronze medal, "was of the reverse-flowered form, carnation in color, marbled and mottled with carmine, the lower petals heavily marked and feathered with carmine-purple; its size, form, and habit were good."

James McTear exhibited 'Jeanie Dean', which was white marked with crimson-purple; other varieties from the same exhibitor were 'Salmonia' and 'Exemplar'. The report of the Massachusetts Horticultural Society for 1864

White gladiolus 'Innocence'.
Photo by Focus no. 5, Shutterstock.com

would indicate that there must have been a remarkable interest in the production of new varieties, for McTear exhibited twelve, Craft thirty-eight, Parkman twenty, and Strong forty-two seedlings during that season. James McTear and George Craft continued to win medals at the shows for the next ten years.

Even Beal, who was closer to that era than we are now, was unable to pinpoint the very first person to introduce a new American-bred cultivar. The best he could say was that Curtis and Cobb's catalogue for the 1868–69 season contained full descriptions of five Craft cultivars and one McTear cultivar. At the annual exhibition held from September 17 to 20 in 1861, McTear exhibited "a spike of gladiolus Calypso, three feet in length, with thirty-two almost perfect flowers." (McTear was in advance of his times and showed the first-ever *Deutzia crenata* at the society in 1866.)

The Washburn catalogues for 1868 contained the first color plate of any gladiolus in North America. James Vick's 'Innocence' was the first American cultivar to be represented by a color plate, in 1885.

At about the same time, the trade began to import Souchet's cultivars from France in enormous quantities. A few years later, in 1893, Luther Burbank introduced his new California cultivars, bred to withstand extremes of climate.

The stems were slightly shorter than the others and much firmer, and the flowers were arranged very closely around them. This made for a very attractive inflorescence. Some flowers were double, though this was not a constant finding. Henry Groff bought all of Burbank's production at the International Show in Buffalo in 1893.

Meanwhile, John Childs released his collection of Childsii cultivars. This series, with very large attractive flowers and vigorous growth, originated with Max Leichtlin. Childs had bought them from a colleague on Long Island, V. Hallock, who had obtained them from the French firm of Godefroy Leboeuf in Argenteuil. Childs then put his own name on them, just as Louis Van Houtte had done with the Bedinghaus cultivars fifty years before. This series was very popular and helped to increase the public's enthusiasm. It is sad to have to chronicle such unscrupulousness in what was otherwise a well-conducted trade.

BREEDERS

Iva Austin (Mrs. A. H. Austin) (1857–1939), Charlestown, Ohio

Mrs. Austin began gardening simply for pleasure, but gradually gathered so many types of gladioli that she decided to start a business in the 1890s. (She was solely known as Mrs. A. H. Austin in all the publications of the period, never using her own first name. That was the rule back then for a respectable married woman.) Her activities were radical for a woman at the time, but a few other women were similarly starting to come into their own. The fact that they dealt in flowers disarmed the Mrs. Grundys of this world since flowers were considered the province of women anyway: fragile, genteel, ornamental, and not really serious. No one thought about the flowers' sexual activities, that the exquisite blossom was really a flaunting invitation to be impregnated. The prudish bishop in Brno entertained the same fallacy about Gregor Mendel's vegetable peas.

In this misty haze few connected the growing and selling of flowers with the realities of running any sort of business, such as paying wages for rough and ready workers, arranging deliveries, and meeting deadlines. None of this came under the rubric of "ladylike."

Mrs. Theodosia Shepherd in Ventura County, California, was one woman who had to earn money rapidly, as her husband was a charming dreamer and the children needed clothes, books, and other necessities. She backed into a profiitable business but ran it very well once she started. Hulda Klager, in Washington State, was obsessed with lilac and made numerous valuable crosses. In England, Hilda Hemus grew sweet peas professionally, employing about twenty men and supplying Samuel Ryder with the seed to fill his penny packages. (For more on this subject see the author's *Visions of Loveliness*.)

Apart from the problem of propriety, there was also the problem for Mrs. Austin that no adequate gladiolus market yet existed. Some of the expansion of commercial gladiolus breeding can be traced to her efforts. She grew her corms on her husband's property, which had been in the family since 1818; Amos Austin had chosen very wisely when he settled in Portage County, Ohio.

As her business expanded, she had to acquire additional land. In its heyday, her farm covered dozens of acres, and she sent corms all over the country. She hybridized many cultivars, firmly discarding all but the ones with the very best possibilities. This rigorous attitude is common to all successful breeders.

The number of corms she grew and distributed was startling, one million in 1911 alone. She used seed from all the known varieties at the time, such as those developed by Lemoine, Kunderd (called Kunderdi), and Luther Burbank. The Gladiolus Society of Ohio elected her vice president, and the American Gladiolus Society put her on its executive committee in 1913. For years she wrote a monthly column in *The Modern Grower,* the gladiolus journal of the day.

B. C. Auten, Carthage, Missouri

Mr. Auten has been rather elusive. Only one reference emerged from a search of the county archives. The 1920 census report listed him as single, aged forty-nine, and in the profession of growing bulbs.

George Burchett (b. 1855), Hampton, Virginia

Burchett emigrated from England in 1872. He settled in Warwick County, Virginia, working in Hampton as a florist and living in Newport News. Very little else is known about him beyond the fact that he left a legacy of about a dozen new gladiolus cultivars.

John Lewis Childs (1856–1921), Floral Park, New York

Childs was a very enterprising and ambitious man, born in Maine. After taking a job with the C. L. Allen nursery in Queens, New York, in 1874, he carefully saved his money and began first renting and then buying as much land as he could in the vicinity. Within five years, his seed catalogue business was said to be very successful.

Once he was established, he became the leader of the town and a state senator for two terms. The name of the town, Floral Park, reflects his powerful influence in that community. Childs was also an enthusiastic bird watcher. He collected hundreds of ornithology books and issued a magazine called *The Warbler.* Another legacy was a small book he wrote, *The Gladiolus,* suggesting that this was an important flower for him.

Arthur Cowee (1859–1939), Berlin, New York

Cowee was a coal merchant by trade but passionate about flowers as an avocation. He dabbled in breeding new kinds of plants in his large garden at home. Cowee particularly liked the gladiolus.

Cowee and Groff met at the Buffalo show and clearly saw that they could work together. Groff employed Cowee as his agent and supplied him with astronomical numbers of corms over the years. Cowee bought land near his home and employed large numbers of men and boys to work in his fields. It was before the days of agricultural mechanization displacing laborers in commercial flower growing.

He developed into a gladiolus grower and breeder in his own right, and, moreover, was a gifted marketer. At the peak of his production, Meadowvale Farms in Berlin was a tourist destination. (This is similar to what later happened in England with Russell's lupines in the 1930s; Russell charged a nominal sum of money for the privilege of gazing at his fields and raised thousands of pounds for the local hospital as a result. See the author's *Visions of Loveliness.*) The coal business was quickly forgotten.

Orders came in from all over North America and around the world, and Cowee became known as the "gladiolus king." It was a measure of his success

Arthur Cowee as a young man.
Reproduced by permission of Sharon B. Klein, Berlin, New York

that he was appointed to the executive committee of the national gladiolus council. One of his brothers, W. J. Cowee, an engineer, designed and made supports for cut flowers and devised solutions for other floral needs, which are still in use today. That business has only just been sold after almost a century in family hands.

Arthur Cowee entered every gladiolus show he could and pushed all his colleagues to do the same. He traveled widely, giving talks to garden clubs and floral societies. His energy and enthusiasm had the desired effect, and sales of corms grew very rapidly. As Alvin Beal at Cornell wrote in 1916, "The popularity of gladioli as garden flowers is due to Mr. Cowee in larger part than to any other person." Groff's varieties were adapted to the climate of southern Canada, and Cowee's to upstate New York, both difficult environments. Cowee could sincerely promote the flowers because he himself was growing them in a place with severe winters.

Berlin, New York, has not forgotten Arthur Cowee. Here is part of an article in the local newspaper, the *Eastwick Press,* from June 2011:

Arthur Cowee's gladiolus fields.
Reproduced by permission of Sharon B. Klein, Berlin, New York

Arthur Cowee's shop.
Reproduced by permission of Sharon B. Klein, Berlin, New York

A bouquet of Arthur Cowee's cultivars. *Reproduced by permission of Sharon B. Klein, Berlin, New York*

Now in its fourth year, the Beautification of Berlin Committee has raised funds to recreate in a small, symbolic way the 200 acres of gladioli planted in Berlin in the late 1800s–early 1900s by Arthur Cowee. Berlin and Cowee were renowned throughout the world as the foremost producers of gladioli bulbs [*sic*] and flowers.

Cowee started with a 40 × 40 plot of gladioli in 1892; by 1905 he had 75 acres of these beautiful flowers and eventually 200 acres were planted. Cowee developed many new varieties over the years, and two were registered in the North American Gladiolus Council, Commercial Growers Division—'Snowbank' which was said, at that time, to be the most beautiful white gladiolus grown, and 'Victory' which was a pale yellow.

Dr. Christian Hoeg (1865–1930), Decorah, Iowa

The end of the nineteenth century saw vast numbers of Scandinavian immigrants entering the United States. Many were oppressed agricultural laborers without much education, but some very highly educated people also came for various reasons. Two of these were Dr. Christian Hoeg, a physician, and Kristian Prestgard, a writer and newspaper editor, both from Norway. Both of them settled in Decorah, Iowa, close to the beginning of the twentieth century and ended up living next door to each other.

Each man became enamored of the gladiolus and bred new varieties in his respective home garden. The gladiolus was coming into its own at the time with men like Henry Groff of Simcoe, Ontario, introducing one amazing cultivar after another. Eventually, Hoeg and Prestgard joined forces and opened the Decorah Gladiolus Gardens, a commercial nursery that lasted until about 1933. Dr. Hoeg died in 1930, but Prestgard continued to run the nursery by himself for another few years.

A volunteer at the Decorah Public Library, Midge Kjome, herself of Norwegian descent, told me she remembers her grandfather buying corms at the nursery. Although the Day family initially founded Decorah after the Ho Chunk (formerly known as Winnebago) tribe was forced from its lands in 1848, numerous Norwegian immigrants arrived in the 1850s, giving it its modern population. This is probably why Hoeg and Prestgard moved there. The town still has a strong Norwegian identity, and Luther College adds extra depth to the community.

Christian Hoeg was born in Bergen, Norway, and studied medicine at the University of Christiania (now Oslo). He immigrated to the United States in 1893, living first in South Dakota, then in Soldier's Grove, Wisconsin, and finally in Decorah, Iowa, where he spent the rest of his life. While in South Dakota, he married Toni Hoeg, a young Norwegian dentist.

Dr. Hoeg was much loved in the community as a conscientious and caring physician. The whole town was aware of his beautiful garden and very proud of his achievements with the gladiolus. Working with Prestgard, editor of the local newspaper *Decorah Posten*, Hoeg introduced many new cultivars. They made the town's name widely known through the Decorah Gladiolus Gardens. The names of Dr. Hoeg's cultivars sometimes echo his Nordic background: 'Jenny Lind' and 'Peer Gynt' are just a few of them.

Kristian Prestgard next to his gladioli.
Reproduced by permission of Vesterheim Norwegian-American Museum, Decorah, Iowa

It is touching to learn that he also collected butterflies. His obituary records his success in that endeavor, too, finding many species not previously known to exist in North America. Several obituaries appeared in the Norwegian-language press of the epoch, as did comments on the life and work of Kristian Prestgard.

Kristian Prestgard (1846–1946), Decorah, Iowa

Prestgard was born in a remote Norwegian village, Heidal i Vaga. He was educated in both Norway and Denmark and worked as a teacher at one time. He switched to journalism and edited a paper in Lillehammer, Norway. Then a group of Scandinavian newspapers sent him to write about the 1893 World's Fair in Chicago, and he never returned to Norway. (Christian Hoeg came to the United States in 1893, and it is tempting to think they may have met at the World's Fair, but that is pure speculation.)

Prestgard moved to Decorah in 1897, the year of his marriage, and took over the *Decorah Posten*. The way he edited the paper brought him considerable prestige, as did the various books he published, and *Decorah Posten* became the most widely read Norwegian-language newspaper in the United

States. For about ten years he also published a Norwegian literary magazine, *Symra*. The king of Norway recognized his achievements by conferring a knighthood on him in 1926. Prestgard was deeply concerned about maintaining ties with Norwegian culture and was a founding member of the Norwegian American Historical Association.

No one will ever know whether Prestgard started breeding gladiolus first and Hoeg followed, or vice versa. Clearly, this was strictly a hobby in both men's cases. Prestgard continued to run the Decorah Gladiolus Gardens after Dr. Hoeg's death in 1930 and only gave it up when his own strength began to wane. Because of his work, he met Henry Wallace, secretary of agriculture and later vice president to Franklin Roosevelt. Some years afterward, Wallace wrote to him and asked him if he was still growing gladioli.

Amos E. Kunderd (1866–1965), Goshen, Indiana

Kunderd bred the first ruffled gladiolus and later introduced the laciniated form. He changed many people's perceptions of this flower. The ruffled 'Kunderdi Glory' appeared in 1903 and had phenomenal success. About twenty years later, he introduced 'Laciniatus', also to great acclaim. Ruffled gladioli are still grown and are widely available as cut flowers.

Part of the excitement stemmed from the fact that very few new types of gladiolus had been introduced for many years. Americans were also pleased that one of their own had caught up to the foreign breeders. One corm of 'Kunderdi Glory' could fetch up to $1,000. Even as late as 1981, sixteen years after Kunderd died, he was one of only two Americans honored in the Gladiolus Hall of Fame at the University of Northern Colorado in Greeley.

Kunderd exemplified the pioneering spirit of the nineteenth century. He was born in a log house near Kendallville and grew up in rural Indiana. Even today, there are only about 9,800 people in Kendallville. His mother grew flowers around their house, and by the age of twelve young Amos was very interested in them. His new flower put him on the map, and a few years later he bought land in Goshen, Indiana, to establish the Gladiolus Farms. He expanded these to two more places in Michigan and at one time farmed more than 750 acres of gladiolus. As he grew older, he switched his attention to the dwarf forms and developed new small cultivars.

Amos Kunderd supplied the rich and the famous with their flowers. He was even invited to the White House to meet President Coolidge, and the president gave him permission to name a new gladiolus after Mrs. Coolidge. The American Gladiolus Society elected him their first president. Everything came to a very sad end with the 1929 crash. The firm went bankrupt, and the business later resumed in a more modest fashion at its original property.

E. E. Stewart (1858–1940), Rives Junction, Michigan

Eugene Elmer Stewart was successful and renowned during his lifetime, but his name has receded into almost complete obscurity. At his peak, he had more than forty acres of gladioli in Jackson County, Michigan. The neighbors and competitors who thought he was crazy when he laid out first five and then ten acres, in the late 1890s, considered that he had lost it completely when he went to twenty-five acres, but when he hit forty acres they started to pay attention.

Stewart was born in Norwalk, Ohio, and moved to Michigan when he was twenty-five. He and his wife, Bertha, had seven children: four sons and three daughters. Whether any of his children took over the business when he died at the age of eighty-two is unknown. His gladiolus business began in his own garden, where his sister helped him plant and harvest the flowers. For more than twenty years, Stewart grew and sold enormous numbers of gladioli, many of them of his own introduction. Two of the better-known cultivars were the light salmon pink 'Fairy', which remained very popular for a long time, and the deep red 'Black Beauty'.

A story in the Jackson County local newspaper about 1917 also mentioned several other favorite cultivars, not necessarily Stewart's own introductions: 'Evelyn Kirkland', 'American', 'Panama', 'Cardinal', 'Fire King', 'Empress of India', 'Florence', and 'Sulphur King'.

Stewart received orders from across the United States and as far away as England and New Zealand. One of the difficulties he had to contend with was a lack of labor. He was active at the time of the First World War, when men were leaving for the forces in large numbers, making it hard for him to get his crops planted and picked.

Almost in desperation, he devised two simple machines that speeded up and mechanized the planting process but required fewer men to operate.

Looking back, one wonders why he did not turn to women to help him. The work was moderately heavy but not beyond the strength of a woman raised in the countryside. He was known for his great honesty and generosity. All Stewart "dozens" actually contained thirteen corms.

GERMANY

Max Leichtlin (1831–1910), Karlsruhe

Leichtlin was an amateur horticulturist only in the sense that he did not depend on his work for his living; but in all other respects, he was a consummate professional. Born into a prosperous family of paper merchants in Karlsruhe, as a youth he was only interested in nature and plants. It was fortunate that his eldest brother took over the family business, because Max could then devote himself to learning horticulture, first at the royal gardens in Potsdam and then in other countries, including South America. For a time he worked with Louis Van Houtte in Ghent.

When his brother died in 1857, Leichtlin had to give this all up and was required to take charge of the family business. The interruption to his chosen avocation ended when his nephew was old enough to handle the business. Then Leichtlin returned to collecting seeds and plants. In his forties, he created a private botanical garden in Baden-Baden and spent the rest of his life working there. Leichtlin left numerous cultivars of several significant species such as orchids and lilies.

He left his mark on gladioli when he used *Gladiolus cruentus* pollen on Gandavensis hybrids. The resulting cultivars were known as 'Leichtilini'. He was attracted to the deep violet-blue cultivar that Lemoine introduced, G. 'Baron J. Hulot'. The color was enriched when Leichtlin crossed this form with other deep blue varieties.

Leichtlin sold his entire stock of cultivars to Messrs Godefroy Leboeuf in Argenteuil, a French firm, in 1882, and they in turn sold them to the Hallock nursery in Queens, New York. A year later, John Lewis Childs, an enterprising and successful nursery owner in New York's Long Island, bought this stock and promptly renamed the cultivars Childsii.

Wilhelm Pfitzer (1821–1905), Fellbach

Pfitzer was born in the village of Fellbach and loved to roam about the countryside as a child. This interest in nature led to him being apprenticed to a nurseryman in nearby Stuttgart. The only thing he wanted to do was work with plants. His father was in the leather business, possibly a saddler, but did not insist on his following the same trade. Wilhelm's parents gave him some property near Stuttgart to start his nursery. Pfitzer built the nursery and laid the basis of its remarkable reputation. His son and grandsons carried it on until after World War II.

The familiar combination of competition from warmer countries and the accumulated stresses of the war led to the Pfitzer nursery closing in the 1980s. Pfitzer's great-grandson, Klaus Pfitzer, has his own small business, primarily raising dahlias.

Most of the Pfitzer gladioli were *Gladiolus* × *gandavensis* crosses. The Pfitzers started out with sturdy cultivars from their own very extensive collection. About ten of their introductions were durable and were exported to the Netherlands, Canada, England, and the United States for a long time. For example, *G.* 'Snow Princess' was still in cultivation in the 1980s.

From a timeline in Bernd Kaiser's monograph on Wilhelm Pfitzer and the nursery, it does not seem that Wilhelm Senior bred many gladioli himself. His sons and grandsons did most of this work. For example, in 1930 the Royal Horticultural Society in London gave the firm an Award of Merit for *Gladiolus* 'Hauptmann Kohl'. By then both Wilhelm Senior and Wilhelm Junior were dead.

THE NETHERLANDS

E. H. Krelage, Haarlem

The Krelage family was a distinguished horticultural dynasty. The first Ernst H. Krelage (1786–1855) founded a nursery in Haarlem in 1811. His son, Jacob Heinrich Krelage (1824–1901), and grandson, Ernst Heinrich Krelage (1869–1956), continued the business. They were particularly famous for their

Gladiole „Reinhold Breitschwert 406"
Original-Aquarelle von G. Ebenhusen, Stuttgart
Gladiole „Neues Jahrhundert 440"

Pfitzer gladiolus cultivars.
Reproduced by permission of Bernd Kaiser

Darwin tulips and their dahlias but were interested in many other bulbs and the broader aspects of botany and horticulture. Jacob Heinrich started a collection of horticultural works, which he donated to Wageningen University in 1916. The nursery won numerous prizes and certificates at the Dutch Horticultural Society shows at the end of the nineteenth century. In 1876, their massing of five hundred plants of *Gladiolus brenchleyensis* at the International Exhibition in Philadelphia was very well received.

Their major contribution in gladiolus was taking the newly introduced *Gladiolus ramosus*, a hybrid despite the specific-sounding name, and backcrossing it with *G. cardinalis*. The resulting flowers were half-hardy early flowering dwarfs, such as G. 'Spitfire', 'Nymph', and 'Blushing Bride'.

Schneevoogt

In the 1880s (though the date is often mistakenly said to be 1838), this Dutch firm introduced what it called *Gladiolus ramosus*, actually a hybrid between *G. cardinalis* and *G. blandus*. The dwarf hybrids were in delicate shades of pink and rose and later became known as the Charm series.

THE STORY has continued into the modern era. Once the catastrophe of World War II was over, breeding gladioli resumed in the European countries, helped by stock that had been safely sequestered in North America and Australasia. Many new groundbreaking cultivars have been introduced, including fragrant ones. Joan Wright of New Zealand succeeded in breeding a reliably fragrant variety in 1955. In Canada, E. F. Palmer brought out 'Picardy', another notable cultivar widely used for further breeding of elegant flowers very much in the way that *G.* × *gandavensis* was so special in its day. Another Canadian, Leonard Butt, made many notable contributions in the twentieth century. Carl Fischer of Minnesota held the record for introducing more than one hundred new cultivars in the mid-twentieth century and winning about forty medals and prizes. He worked with Ralph Baerman, who had bought hybridizing stock from Kristian Prestgard in Decorah.

5

Dianthus

(Carnations and Pinks)

THE STORY of intentional crossbreeding actually started with carnations. Carnations and pinks are as much beloved for their fragrance as for their appearance. Almost three hundred years ago, a nurseryman in London, Thomas Fairchild, had the idea of crossing a pink, *Dianthus caryophyllus,* with a Sweet William, *Dianthus barbatus,* to see what would happen. What happened changed horticulture. Fairchild was simultaneously elated and terrified. On the one hand, he had created a new flower. On the other hand, he had usurped God's powers. Creation was considered solely to be in God's province. Fairchild never dared to do this again.

These flowers go back very far into English and European history, being commonly traced to Theophrastus, the fourth-century BCE Greek naturalist. He gave the flower its generic name based on the words *dios* (god), and *anthos* (flower), elided into *Dianthus*. Linnaeus used this name when he came to classify the plants about two thousand years later.

The name "carnation" can be attributed to several possible sources. One is its connection to the Latin word for flesh, *carne,* since the original species

Dianthus caryophyllus.
Photo by peeravit, Shutterstock.com

flowers were a soft pink color. Another suggestion is that it is a corruption of *coronation,* based on the arrangement of the florets in little crowns or coronets. The name "pink" probably arises from the deckled edges of the petals. A type of scissors called "pinking shears" is used to give fabric the same deckled edge. Using the word *pink* to describe a color seems to have happened after the plant became widespread. Until the late seventeenth century, there was no word to describe the color now known as "pink." It was called "light red."

BOTANY

There are about three hundred species of dianthus, largely found in eastern, central, and southern Europe, with some species occurring in China and Japan. One prostrate species, *Dianthus repens,* is found above the Arctic Circle in North America, and there is a single species endemic to northern Africa. Different classifications are based on a variety of characteristics.

Botanical classification follows standard scientific convention, but there is also a horticultural version and a floricultural one, based on the anatomy of the blossoms. Carnations can also be grouped by their colors. The distinction between a pink and a carnation is blurred, depending largely on history and usage. Botanically, they are all closely related.

The horticultural divisions are as follows:

Border carnations
Perpetual-flowering carnations
    Malmaison carnations (highly fragrant with very large flowers)
Border pinks
    Antique border pinks, introduced before 1920
    Modern border pinks
Rock garden pinks
China pinks and their hybrids
Clusterhead pinks
Species pinks

Within the border carnations, there are the self, fancy, and picotee divisions, based on the color and shape of the petals.

A few of the botanical species, among them *Dianthus caryophyllus, D. barbatus, D. chinensis,* and *D. plumarius,* have been used in gardening and commerce for centuries. Spontaneous hybridization has not been widely reported among the species, but they can be intentionally crossbred. This capacity is the basis of the enormous number of modern hybrids.

Charming references to carnations, pinks, gillyflowers, and other folk names such as "sops in wine" may be found in sixteenth- and seventeenth-century works such as John Gerard's *Herball,* John Parkinson's *Paradisi in Sole,* and William Turner's *New Herball.* The strong clove-like fragrance led to the flower being used to season indifferent wines or dull-flavored food, because it was a good deal cheaper than imported cloves from Africa.

COMMERCIAL ASPECTS

Carnations are very big business in modern floriculture. In each of the decennial horticultural censuses of the U.S. Department of Agriculture (USDA) for the past fifty years, the department has counted them as bunches of cut flowers. Three states, California, Florida, and Michigan, accounted for the largest numbers of carnations.

The aggregate income from carnations grown in the United States has been shrinking recently. Since the flowers remain very popular and widely available, this reduction in gross sales may simply reflect the increase in imported flowers from Central and South America, not counted in the USDA census. This is borne out by a recent story in the California press about the closure of many carnation farms in and around Salinas within the past few years, as imports take the place of their crops.

The flowers' huge range of colors and shapes as well as their fragrance keep them popular. Finding out how they came to have this range of color and shape is the substance of the next section.

HYBRIDIZATION

Thomas Fairchild performed the first known modern intentional crossbreeding between any two species of flower in 1720, using *Dianthus caryophyllus* and *D. barbatus*. He was successful, as mentioned at the beginning of this chapter, but the achievement filled him with religious trepidation. For years afterward this flower was known as "Fairchild's mule."

What follows is really only a rather brief discussion of a large subject, touching on the highlights. Trying to cover all the byways of carnation and pink history in the British Isles alone would be impossible. Many species of dianthus do very well in England's climate if they are protected over the winter by a cold frame or greenhouse. They also flourish in pots.

The florists (see chapter 2) were very enthusiastic about certain types of dianthus, especially the laced pink, white with a black center. The pattern of the two colors offered them numerous challenges, though the results were not necessarily pleasing to our eyes. They used rigorous selection to achieve their goals of breeding championship flowers with rather bizarre specifications, such as a totally circular face, but they do not seem to have picked up on Fairchild's ideas to make their own crosses. The fragile and complicated pinks only bloomed once in the summer, and the florists were not interested in doing anything else about them.

Thomas Fairchild's "mule." *Reproduced by permission of University of Oxford Herbarium*

Everything changed in the world of dianthus when two gardeners in Lyons, Dalmais and Schmitt, introduced new cultivars between 1840 and 1866. The National Carnation and Picotee Society of England was founded in 1851, presumably because of this impetus. *Dianthus* had become known in America at some time in the late 1820s. A carnation seedling was shown at a meeting of the Massachusetts Horticultural Society in 1829 and in 1831 the society offered a prize for the flower. This track petered out, and the true origin of modern American dianthus can be traced to the new French cultivars mentioned above. Here were perpetually flowering plants that could be grown successfully out in the open and did not need coddling.

Monsieur Dalmais of Lyon, a well-known amateur gardener, registered the results of his breeding the remontant cultivar *Dianthus* 'De Mahon' with *D.* 'Biohon' (or 'Bielson', according to Charles Ward) and then crossing the F1 generation with pollen from a Flemish cultivar in 1842. According to an 1886

59 (above). *Two pinks raised by Dr Allan Maclean of Colchester, 'Miss Eaton' and 'Miss Nightingale' (the 'Florist', 1857). These nineteenth-century illustrations in horticultural journals were very much idealised.* (Courtesy of Mrs Susan Farquhar.)

60 (top right). *'John Ball', a pink raised by Dr Allan Maclean in the 1850s, the only one of his to survive to the present day.*

61 (second from top). *The pink 'Beauty of Healey', raised by W. Grindrod before 1925.*

62 (third from top). *The very dark laced pink, 'Old Velvet'.*

63 (bottom). *The pink 'Laura Jane', raised by S. Webb of Oxford in 1980 and photographed by him.*

Florists' pinks.
Source: The late Ruth Duthie

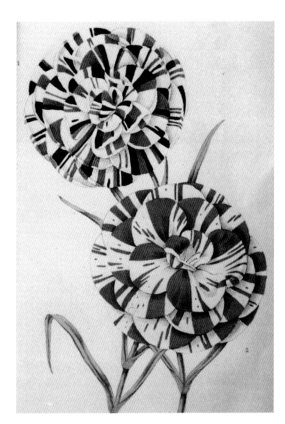

Florists' carnations. *Source: The late Ruth Duthie*

American Florist article, Dalmais kept selecting improved plants from further crosses until the progeny grew true and he could safely put them out into the market. Two years later, he showed many more cultivars with a far wider range of colors. The only disadvantage was that all these plants were rather tall and leggy, leading them to be called "tree carnations." *D*. 'Atim' remained the best known.

A colleague of Dalmais's, a Monsieur Schmitt, took up the work and introduced more new cultivars some years later. This work culminated in that of another Lyonnais breeder, Alphonse Alegatière. In 1856, Alegatière introduced the cultivars 'Edwardsii' and 'La Pureté', which remained significant for many years.

Alegatière's vastly improved versions were taken up avidly by English growers and used to supply carnations for the cut flower trade at Covent Garden. One way to encourage large clusters of bloom was to prevent the new plants from flowering for two years. Growers also trained carnations up the greenhouse roofs. The Malmaison series was a subset of this group with exceptionally large blossoms, as much as six inches in some cases. They were used by other breeders to create additional variations on a very successful theme. The stems were shorter and stronger, the petals were tough, and the blossoming continued for up to five months. Levi Lamborn considered that they were the foundation of all modern American carnations.

The French nurseryman Charles Marc was the first person to take the French cultivars to America, in 1852. He offered them for sale in Flatbush (now part of Brooklyn). Other European firms also set up shop in Brooklyn in that epoch. Perhaps the best known was Dailledouze, Gard and Zeller. In 1866 this company introduced more than fifty new cultivars of dianthus.

Charles Zeller later said that he and his partners, John Dailledouze and Joseph Gard, obtained remontant dianthus seed from Lyon in 1858, as well as some plants of Alegatière's 'La Pureté' in deep rose, together with the white 'Mont Blanc' and 'Manteaux Royal', variegated red and white. The first plant they showed as a result of their crosses was fringed and white, the very floriferous 'Mrs. Degraw', named for the wife of the president of the Brooklyn Horticultural Society. Another pure white cultivar, 'Flatbush', appeared in 1864.

This information is borne out by an undated catalogue they issued at some point between 1862 and 1872 listing fifty-three cultivars of *Dianthus*, according to Charles Ward. So far no copy of it has been found, but Ward is reliable. He was active only a few years after this took place, managing the Cottage Gardens nursery in Queens, New York.

Dailledouze, Gard and Zeller actually hybridized some of these new cultivars, but the rest were other people's work. While carnations are almost universally available now, at that time very few florists of the 112 firms listed in the 1860 census bothered to pick up this obscure European flower. It was a French niche in New York until 1875, when Charles Starr of Avondale in Pennsylvania introduced the first of his long series of cultivars, 'Lady Emma'. According to Lamborn, Starr eventually introduced fifty new cultivars, including 'Chester Pride' (1877) and 'Buttercup' (1878), and made the flower very popular with the rest of America. Starr also commissioned the earliest known engraving of a carnation in America, used to illustrate Lamborn's book, *American Carnation Culture*, in 1885. Starr died in 1891.

Lamborn was a very good observer and kept excellent records. Looking back almost forty years to 1897, he commented that 17 men had each contributed between five and fifty-five new cultivars in the United States, accounting for the bulk of the new introductions; 30 had contributed between one and five new cultivars; and 127 had left only one each. He added that of the roughly eight hundred cultivars he knew about, perhaps fifty were truly new and important additions to the floral trade. The leaders in his tally were Charles Starr, Frederick Dorner, and W. Simmons.

To show how evanescent these introductions were, Lamborn listed about fifty of the plants most favored by the early growers and noted that within fifteen years not one of them was grown anymore. He offered the very sage advice that growers should observe how their new plants performed for at least five years before putting them on the market. If only James Hartshorn had done this with 'Fiancée' (see below), he would have saved himself a lot of anguish.

In contrast, Lamborn emphasized the very pivotal role played by Charles Starr and his cultivar 'Lady Emma'. In the 1870s, Starr used 'La Pureté', 'Edwardsii', and 'Astoria' as parents to improve the structure of the blossom. 'Astoria' was another descendant of Alegatière's introductions. It was bred by a Mr. Wilson of Astoria in New York and was said to be the first carnation of its type grown in the United States.

Starr found it was that was easier to change the structure of a flower than to improve its color or fragrance. Color was particularly fickle and difficult to pin down. One grower complained that he crossed two crimson flowers and came up with white offspring. Although Gregor Mendel had published his extraordinary results in 1865, in an obscure Czech journal, scientists did not know about his work until 1899 when Bateson, Czermak, and others discovered it. Without that knowledge flower breeders were still working in the dark.

Lamborn also admired Richard Witterstaetter in Cincinnati, considering him to be "reliable and painstaking." Witterstaetter introduced many new cultivars. Skidelsky also had a high opinion of Witterstaetter, finding him taciturn but very upright and straightforward.

At the same time all this was happening, the carnation appears to have been developing independently in California. Victoria Padilla wrote an excellent book on the gardens of southern California in 1961. In it she suggests that carnations reached California directly very early and were not just imported from the East Coast. She referred to the antique agricultural textbook by Alonso de Herrera, found in the library of the Franciscan mission at Santa Barbara and carefully annotated by Fray Antonio Jayme in 1797. In the section on ornamental flowers, Herrera mentioned carnations. Much of what the missionaries brought with them from their predecessors in Mexico was subsequently grown in California, though the Franciscans kept no records of such trivial and profane things as fruit trees and flowers.

Padilla also believed that the Spanish soldiers' wives took seeds of precious and fragrant plants from home when they embarked on the long and perilous voyage to a place that was then even more isolated than Siberia. Mid-nineteenth-century visitors to the California missions left tantalizing accounts of seeing gardens full of flowers even though the missionaries had been dispossessed by the Mexican government, but these travelers do not say exactly which flowers they saw.

California nurserymen saw great potential in carnations as cut flowers and planted them in great numbers to supply this market. The German nurseryman Franz Hosp moved from Cincinnati to Riverside, California, in 1888 to open his business. He devoted fifteen acres to carnations, though which variety he grew is unknown. Although he was the first to do this, he did not have the largest grounds in the state: E. J. Vawter grew carnations on forty acres near Santa Monica. This forms the background to the activities of Stephen Lenton, a grower and hybridizer who went to Ventura County in 1888 when it was still a dusty backwater (see below).

INDIVIDUAL CARNATION BREEDERS

ENGLAND

In 1868, soon after the French nurserymen in New York had become established, Charles Turner of the Royal Nursery, Slough, introduced the striking fringed white carnation, 'Mrs. Sinkins', which was a sensation in its day and is still available for sale on the Internet.

I discussed Turner a bit in the chapter on dahlias in *Visions of Loveliness*. Curiously, although nowadays we remember certain names in association with particular flowers, during their lifetimes these growers were not so specialized. Turner bred prizewinning dahlias, for example, even though he is now associated with dianthus and roses. As another example, John Keynes (grandfather of the economist John Maynard Keynes) is associated with the dahlia, but he also bred prizewinning carnations. One has to remember that these men were first and foremost businessmen making a living. If something were fashionable, they quickly pursued it to take advantage of the brief surge in business. Many took up carnations at that time for that reason.

Allwood Brothers

Montague Allwood (1880–1958) became the best known of the three brothers. In 1910, his elder brother Edward put together the money that allowed them to buy land and develop their nursery. The middle brother, George, had worked in carnation production in the United States and was technically skilled. Montague himself had started out by washing pots in a local nursery and had risen to manage carnation production. The owner offered him a partnership, but he preferred to go out on his own with his brothers.

They came from a Lincolnshire farming family, but land in Lincolnshire was too expensive, so they moved to Sussex, near the south coast. There were other advantages too. The climate and light were moderately better, though the heavy clay soil did need constant attention. Even in an island like England, only four hundred miles from top to bottom, there is a temperature gradient from north to south. Robert Bolton had moved his sweet pea nursery from Northumberland to Essex for the same reason.

Montague was personable, outgoing, and a splendid promoter. At the Chelsea flower shows, he always presented the Queen with a small posy, and sent buttonholes for the royal men. The shows were held in an enormous canvas marquee, which obscured much of the light. Without the warmth of the sun, pinks lose their fragrance, so Montague got busy and created a carnation perfume for his staff to wear.

Dianthus × *allwoodii* 'Doris'.
From Fred C. Smith, A Plantsman's Guide to Carnations and Pinks *(London: Ward Lock, 1990). Attempts at tracing the copyright holder of this image through the Orion Publishing Group were unsuccessful.*

He also published a popular book about carnation culture, which went through several editions. Despite all these distractions, he managed to introduce several new cultivars of dianthus, *D.* × *allwoodii*. Perhaps the best known of them is 'Doris', a double pink blossom with darker salmon pink markings, introduced in 1945 and still available today. Doris was Mrs. Montague Allwood's first name.

The *D. × allwoodii* series was hardy with a rich fragrance and flowered for a long time. It was a cross between the perpetual flowering carnation, *D. caryophyllus*, and *D. plumarius*. The Scientific Committee of the Royal Horticultural Society gave the race its name.

The current owners of the business, David and Emma Jones, tell the story that "Mont," as he was always called, was not too keen on 'Doris' when it first appeared and instructed his staff to discard it, but the foreman disagreed and quietly propagated it in a remote corner of the nursery. When Mont took another look at it later, he saw its quality, and they moved ahead with production.

James Douglas, The House of Douglas, Eden Nursery at Great Bookham, Surrey

James Douglas was born in 1837 in the village of Ednam, Roxburghshire, Scotland, and, like so many before him, went south in 1860 to make his fortune. From the age of fourteen, he held responsible jobs as a practical gardener on several large Scottish estates, first walking miles each way from home and later living in the rather primitive bothies. In spite of that, he still found the time and energy to read and learn. Nothing he ever learned was lost on him. When the head gardener at his first job taught him how to pollinate flowers to breed new ones he remembered it long afterward. This varied experience clearly influenced James Veitch to take him on in London at the Chelsea establishment.

Veitch quickly entrusted him with more serious work, and he began to help lay out the firm's exhibits at a Royal Horticultural Society's flower show at its new gardens in South Kensington. Douglas's talent and skill were bound to impress others, and he left Veitch to go into private service, spending thirty-four years with Francis Whitbourn in Ilford, Essex. This was essentially a lifetime, during which he married and had five children.

Whitbourn, an enlightened employer, encouraged Douglas to enter competitions and win prizes whenever he could. Whitbourn was the beneficiary, of course. He could hold his head high in local society with a prizewinning gardener reflecting glory on his estate. (Whenever one reads of this or that aristocrat introducing new flowers, the work was actually done by the gardener.) Douglas must have had boundless energy, for not only did he win all the prizes for fruit and vegetables but also he started to hybridize dianthus and many other ornamental plants.

At the same time, he taught Sunday school at the Congregationalist church in nearby Barking. He also became an examiner for the Royal Horticultural Society's educational programs in the London boroughs and other parts of the country. In 1876, he was a founding member of the National Carnation Society, and a few months later he helped to found the National Auricula and Primula Society.

The carnation of the time needed a stronger stem and a better range of colors to become truly popular. In 1889, Douglas went to Germany with Martin Smith of Hayes in Kent, a prominent amateur breeder, seeking good breeding stock with which to start these improvements. Ernst Benary of Erfurt provided about two dozen varieties, which turned out to be very fruitful. The yellow ones were an exciting innovation. Eventually, Smith and Douglas introduced almost two hundred new cultivars. A few of them are still in commerce.

Whitbourn died in the 1880s but left a provision in his will to keep Douglas on as head gardener. About five years later, Douglas bought land in Surrey and started his own commercial enterprise with carnations and auriculas. He had first seen edged auriculas at Edenside (an estate owned by James Tait). The flowers made a deep impression on him, and Douglas took the name of his nursery from that estate. The firm, Edenside Nursery, began in 1893. By 1897 the nursery was sufficiently successful for him to leave Mrs. Whitbourn's service and concentrate on his own business. Douglas shared a great deal of what he had learned over the years with the interested public through articles in the horticultural literature and a short book, *Hard Florists' Flowers* (1879).

Real recognition came with the award of one of the very first Victoria Medals of Honour by the Royal Horticultural Society in 1899. Douglas remained remarkably active but died suddenly after an operation in 1911. He left a substantial estate for the time. His son James and grandson Gordon both made heroic efforts to continue the business despite the two subsequent world wars. Eventually they had to face reality and give up.

The municipality took over much of the land for housing. As a nod to the nursery, some of the streets were named after long-serving employees of

Edenside. When Gordon Douglas ceased running the business, he still sold some flowers from his own garden. The rest of the stock was said to have gone to the Allwood Brothers.

Carl G. Engelmann (1874–1941), Saffron Walden, Essex

Engelmann is essentially unknown now, but at least twenty-six cultivars are attributed to him. His grandson and great-granddaughter very graciously shared an article about his work with me that appeared in a Saffron Walden historical magazine a few years ago.

Carl Gustav Engelmann was born in Germany in 1874 and moved to England around 1895 after a thorough training at a horticultural college, the sort of training that was unique to Germany at that time. His family were nurserymen and had been in the Grand Duchy of Anhalt for many years. The state of Saxony-Anhalt is in a very fertile part of Germany. Carl went to work for Uzzells, a nursery near Hampton, along the River Thames, but after only a couple of years decided to open his own business in Saffron Walden, a small town in Essex.

This was an agricultural district and had the right kind of slightly chalky soil that is good for carnations. Engelmann had bought five fields, which were later covered in greenhouses. (Perhaps marrying Miss Charlotte Amelia Uzzell had something to do with being able to afford this land.) The greenhouses were devoted to growing carnations. They had to be heated by coal, but a plentiful supply of this fuel was available.

Increasing prosperity in the Victorian middle class allowed the public to indulge lavishly in cut flowers for their houses, and Carl Engelmann was very good at reading the signs. A luxury market is fickle and constantly changing, but fortunately carnations remained in favor for a very long time. Engelmann worked with a long-stemmed perpetual cultivar, or "tree carnation," which was originally a French breakthrough but had made the carnation highly successful in America. He was growing 'Enchantress' (developed by Peter Fisher of Ellis, Massachusetts, in 1903) before 1913, and his was perhaps the first nursery to do that on a large scale in England.

Nothing is reported in this article about the effects of the First World War on Engelmann and his business, but evidently he weathered it all right. Many perfectly loyal people in the United Kingdom who came from Germany were vilified, and their premises were vandalized by an ignorant and hysterical public. Even the royal family had to adopt an English name, House of Windsor, to replace the German Saxe-Coburg and Gotha, acquired when Queen Victoria had married Prince Albert of Saxe-Coburg and Gotha. The current Queen Elizabeth's uncle's family name of Battenberg became Mountbatten.

Everything took a great deal of hard work, and Mrs. Engelmann worked closely with her husband in packing the flowers for shipping. It later became possible to construct a special packing house and employ extra people to do this job.

Carl Engelmann played his part in civic affairs as a member of the borough council. He also served on a committee at the Royal Horticultural Society and gained some national attention. In 1925, the Engelmann nursery was awarded the Sherwood Cup at the Chelsea Flower Show.

At some time during these productive years, Engelmann also began hybridizing his own carnations. Between 1909 and 1919, he introduced new cultivars bred from a wide range of parents. The nursery also offered many other standard garden plants and was known for its giant pansies. After Carl's death in 1941, his younger son, Eric, ran the firm for more than thirty years. It had to close in 1975 when the cost of oil rose by 400 percent and many other factors conspired against them.

Simon Low and Co., Bush Hill Park, near London

Although the Lows were mainly known for their orchids, they also introduced twenty-four cultivars of carnation.

Charles Turner, The Royal Nursery, Slough

Turner was a very energetic and effective man. He started in a small village in Berkshire and then took over E. and C. Brown's dahlia nursery in Slough,

Dianthus 'Mrs. Sinkins'.
Courtesy of the Royal Horticultural Society

Buckinghamshire. He was the principal force behind the founding of the National Carnation and Picotee Society in 1851. Even as interest in dahlias died down, he turned to the carnation, at just at the right time.

Slough holds him dear for the introduction of *Dianthus* 'Mrs. Sinkins'. The master of the Slough Workhouse in the 1860s was a Mr. Sinkins. He enjoyed dabbling in his garden when not chivvying the unfortunate inmates of his establishment. One summer he found a gorgeous fragrant, fringed and double white carnation growing there and named it for his wife. Flower or no flower, somehow Charles Dickens's "Mr. Bumble" and Oliver Twist don't seem very far away. Sinkins handed the flower over to Charles Turner to propagate and it became very successful.

Turner had a fine reputation, and Charles Darwin wrote to him in 1863 with a question about the spacing of newly crossed hollyhocks. Charles Darwin had seen bees pollinating these flowers and was concerned that unless the experimental hollyhocks were separated widely enough, this would ruin any attempt to breed a new pure hybrid. Turner's actual reply has not been found, but Darwin incorporated it into his book about the variations among plants and animals.

Channel Islands

H. Burnett, Forrest Road, Guernsey

Burnett won a Silver Flora Medal at the Royal Horticultural Society for his carnations in 1907. The register lists twenty-four of his cultivars. Allwood admired him for the introduction of 'Mrs. H. Burnett', a most useful cultivar and parent to many other excellent perpetual flowering carnations. Burnett obtained it by crossing 'Mrs. W. T. Lawson' with 'Pride of Market' and released it in 1903.

FRANCE

Alphonse Alegatière (1821–1893), Lyon

Monsieur Alegatière gave up his leather tanning business because of a chronic illness and took up hybridizing carnations instead. He introduced many cultivars with good stiff stems, including dwarf forms. Alegatière also showed that layering the plants to propagate them was unnecessary, because cuttings worked very well.

Alegatière's obituary in *Lyon horticole* recounted his use of numerous different parents in his efforts to obtain better flowers, such as the *mignardise* (pinks), Chinese carnation, "poets' carnations" *(Dianthus barbatus),* and remontant carnations already in commerce. The obituary described his imperturbability in the face of frequent failure, as he remained certain that he would succeed sooner or later. One of his most valuable results was a plant with stems so strong that they no longer needed ungainly staking ("éliminer les tuteurs au bois, si disgracieux"). Two of his most successful cultivars were

'La Pureté' and 'Edwardsii'. 'La Pureté' came in pink, white, and red and was used in breeding more flowers for many years.

The obituarist was outraged that the French government never gave Alegatière any recognition or financial assistance as he grew older. "V-M" thought he should at least have received the Livre D'or du Mérite Agricole.

Interestingly, once Alegatière succeeded with carnations, he turned to improving roses and pelargoniums instead. In that way, he resembled the Canadian breeder Henry Groff (see chapter 4). In addition to his other qualities, Alegatière was an inspired teacher. Young Joseph Duchet, a future star in the rose breeding world, went to work under him to learn as much as he could about roses.

Monsieur Dalmais, Lyon

Dalmais worked for a well-known amateur, one Monsieur Lacène, a founder of Lyon Horticole. Dalmais crossed 'Mahon' with 'Bielson' and obtained 'Atim' in 1842. (The names of these cultivars have been spelled differently over the years. These are the standard spellings today.) He released 'Atim' in 1844 and then bred more cultivars in different colors. The flowers bloomed repeatedly but were rather tall and leggy. The English called them "tree carnations."

Monsieur Schmitt, Lyon

Schmitt was a professional horticulturist, roughly contemporary with Dalmais. He introduced several excellent cultivars, such as 'Arc en Ciel', but all his plants succumbed to disease and he gave the work up.

UNITED STATES

Adolphe Baur (1876–1942), Indianapolis, Indiana

Adolphe Baur was born in Pennsylvania in 1876 and moved to Indianapolis after obtaining horticultural training in Louisville, Kentucky. He started his own floral business in 1899 with his partner, F. S. Smith. The partnership dissolved in 1911, and he took in a new partner, O. E. Steinkamp. They incorporated the firm in 1922 with Edward Larsen, and that is how it remained at the time of Baur's death in 1942. The firm persisted until 1953, when it finally closed for good.

Baur was noted for his carnations, but he was also a leading figure in the chrysanthemum world. He was president of the American Carnation Society, the Chrysanthemum Society of America, and the Indiana State Florist Association at various times. Baur and Steinkamp introduced more than twenty cultivars of dianthus, some based on 'Armazindy', with names like 'Indianapolis' and 'Indiana Markey'.

His son, Francis Baur, succeeded him and became president of the American Carnation Society in 1954, ultimately winning four gold medals for new carnations. Francis was a graduate of Purdue University and died in 1989.

Chicago Carnation Company, Joliet, Illinois

The Chicago Carnation Company was based in Joliet, Illinois, about fifty miles from Chicago. The company began when a native of Joliet, James Higinbotham, employed James Hartshorn to be the gardener for both his property in Joliet and that in Chicago in 1897. They shared an interest in carnations, which were coming into their own at the time. Higinbotham appointed Hartshorn to manage the new company. In 1904 they had premises at 1001 Cass Street, in Joliet.

Quite soon afterward, in 1901, another carnation company, the Joliet Floral Company, appeared in Joliet, belonging to J. D. Thompson. Simon Skidelsky, a traveling salesman for a firm of horticultural suppliers, called on the Chicago Carnation Company as part of his commercial rounds and got to know Hartshorn quite well. Both companies persisted at least until 1930. The Chicago Carnation Company was big enough for the railroad to run a separate spur onto their property to ship the flowers. Both companies maintained very large greenhouses off premises.

The Chicago Carnation Company registered about twenty-five excellent cultivars, among them 'Mrs. N. M. Higinbotham', 'Crusader', and 'Harlowarden', which was used to breed other cultivars in its turn. They registered their last one in 1916: 'Aviator'. This does not mean they did not introduce other flowers, simply that these were not in the registry.

One cultivar, 'Fiancée', did not fare so well. Lamborn put it somewhat lugubriously for this and other carnations in the same plight: "They should quietly sleep in the catacombs of defunct carnations." Skidelsky recorded the sad story in his book, *The Tales of a Traveler*. It was the classic tale of chickens not hatching according to plan. 'Fiancée' was going to be the sensation of the year, making everyone's fortune, but things did not work out the way they were supposed to do.

'Fiancée' was bred by Frederick Dorner, the founder of a large and important nursery in Indiana, but then sold for $15,000 (a fortune in those days) to the Chicago Carnation Company in 1905 for propagation. Skidelsky thought Dorner had some reservations about taking this money, but even he could not resist it. Hartshorn was "as gleeful as a ten-year-old" over the flower's prospects. A vase of "this most magnificent flower" had been taken to the Chicago show. "Its appearance created a sensation among the Carnation growers, the like of which had never been known before," but even the tough flower breeders of Chicago were startled when Hartshorn undertook to issue half a million 'Fiancée' plants.

Skidelsky promoted 'Fiancée' tirelessly and in short time had an order for 10,000 cuttings. Orders simply poured in, and would be doubled and tripled before they could even be entered in the order books. Skidelsky visited Joliet once or twice in the summer to be sure everything was going well. Hartshorn was in "high spirits," but the following January when the seedlings should have been ready, disaster struck.

Hartshorn had failed to get the cuttings to root, no matter what he tried. He had no stock to fulfill the orders. 'Fiancée' was not a wonder plant at all, but a very questionable one that should never have been sent to the market. Skidelsky rushed back to Joliet and wanted to take Mr. Hartshorn to task, "but poor man, he looked so forlorn and dejected," Skidelsky did not have the heart to berate him. Hartshorn was doing it to himself.

Skidelsky salvaged what few viable plants there were for his customers, but after looking through dozens of angry letters Hartshorn threw the letters aside and asked Skidelsky "to accompany him to his club." There they spent the rest of the day eating sandwiches and drinking shots. Six months later Hartshorn was dead, following complications of a surgical operation. There were those who thought he had lost the will to live. The name 'Fiancée' is perilously close to *fiasco*.

Cottage Gardens Company, Queens, New York

The manager of this nursery was Charles Willis Ward, author of a very useful book about the American carnation. Among other things, he gives brief biographies of the principal carnation figures who were active at the time he wrote it, circa 1911. He and his staff introduced more than twenty-seven new cultivars.

George Creighton, New Hamburgh, New York

Creighton was not as active as some of the other breeders, but he did leave a legacy of eight cultivars.

Dailledouze and Zeller, Brooklyn, New York

This firm was responsible for more than fifty new cultivars, starting in 1866 (see earlier discussion). Charles Ward left a detailed account of the way they began their business. John Dailledouze was French while Charles Zeller was Swiss.

Frederick Dorner, Lafayette, Indiana

Dorner is discussed elsewhere in this book (see chapter 2). He introduced numerous new cultivars of carnation.

John Dailledouze.
From Charles Willis Ward: The American Carnation *(1903).*
Reproduced compliments of Applewood Books, Carlisle, MA

Charles Zeller.
Reproduced compliments of Applewood Books

Frederick Dorner.
Reproduced compliments of Applewood Books

Henry Eichholz, Waynesboro, Pennsylvania

Eichholz was an Austrian immigrant born in 1864 who started his Pennsylvania nursery in 1894 in a rented greenhouse. A few years later, he managed to buy his own property: fourteen acres at 9 State Hill Road. The greenhouses were closed in 1935, and his son moved to other premises. The firm introduced twelve cultivars.

Peter Fisher (b. 1857), Ellis, Massachusetts

Fisher was born in Scotland in 1857 on the estate of the Duke of Athol, in the village of Dunkeld, Perthshire. His father was the duke's land agent. When Peter was fifteen, he began his gardening apprenticeship on the dowager duchess's estate nearby. Like so many other Scottish gardening apprentices, he moved to London to obtain broader training and experience.

Clearly he was ambitious and very able. In 1884, he and one of his brothers immigrated to the United States and began working as gardeners on various Massachusetts estates. Within a couple of years, they had gone out on their own as the Fisher Brothers Company. Several years later, Peter sold his share of the firm to his brother and started his own nursery in Ellis.

He took up the very fashionable and lucrative carnation business in earnest. Quite quickly, he introduced remarkable new cultivars, thirteen in total, some of which influenced future breeding for a long time. In 1892, there was 'Mrs. Lawson'. That was followed by 'Enchantress', which gave rise to delightful sports in pink and red. Fisher seemed very adept at public relations, and several newspaper articles extolled his new varieties.

Peter Fisher's brother may have been A. Sewall Fisher, who worked as a florist in Warren Court in Framingham between 1884 and 1902. The city directory of the latter date noted that he had moved to Boston.

Peter Henderson (1822–1890), New York

Henderson was a leading figure in nineteenth-century American horticulture. As was the case with so many other gardeners and nurserymen, he emigrated from Scotland (where his maternal grandfather, Peter Gilchrist, had started life as a shepherd but rose to become the owner of a thriving nursery). In 1843, after a four-year apprenticeship at Melville Castle, Henderson arrived in New York. He was legendary for being enormously industrious, publishing several books in addition to running a huge nursery business. Another quality separated him from many business proprietors: sensitivity to the feelings of a young man looking for a job. He himself had been so rudely turned away when he arrived in New York and asked for a job that he resolved never to do that to anyone else.

George Thorburn in Queens, a well-known and successful nurseryman, was the one who had finally taken a chance on the newcomer. A year later, Henderson moved to Philadelphia to work for Robert Buist, another Scot. Buist played an important role in the early distribution of the poinsettia (see chapter 1).

In 1847, Henderson began a nursery in Jersey City in partnership with his brother James. They stayed together for a few years but then split up. James bought another property where he grew and sold vegetables, but Peter thought he could do better with ornamental plants and stayed on at the original place. As part of this process, he added more and more greenhouses to his lot and soon had to buy additional land. Twenty years later, after many expansions and changes, he established the final presence by which he is remembered today: Peter Henderson, Seed Merchant, in Cortland Street, New York City.

Because of his skill and generosity, he became the acknowledged leader of horticulture both in New York and in the rest of the country. Many of the modifications he invented to simplify and improve his own business were eagerly adopted by other nurserymen once he pioneered them. One valuable idea was to offer everything someone might need for the garden in one shop, a revolutionary notion in its day.

When Mrs. Theodosia Shepherd in Ventura, California, developed a new line of petunias, her first thought was to send them to Henderson and get his imprimatur. That encouraged her to go ahead and work on her packaged seed business. Henderson believed that if you could read you could succeed at almost anything you wanted to do, and horticulture was no exception. He demystified the processes needed to establish a successful nursery. His ability to write simply and articulately was another component of his success. Besides six books, Henderson wrote all his own catalogues and advertisements.

There were very few types of flower with which he was not associated. He was able to discern which novelties would be useful and have staying power better than many of his contemporaries. His son Alfred wrote a memoir about his father in 1890, giving a general overview of his accomplishments.

One of Alfred's observations about his father's first book is quite unexpected: Alfred believed that *Gardening for Profit,* published in 1866, helped in the reconstruction of the South after the Civil War. He claimed it enabled people to start their own businesses while normal life was still confused and disrupted. He did not give any evidence for this amazing statement, but it is intriguing.

To start his carnation business, Henderson bought fifty seedlings in five-inch pots from Dailledouze and Zeller in 1864 for $1.50 each, possibly the first wholesale purchase of the flowers in the United States.

Stephen Lenton (1847–1905), Piru City, California

Lenton had been born near Birmingham in England and immigrated to the United States in 1873, bringing his wife and infant son, Albert, with him. Lenton was a practicing Christian in Elgin, Illinois, and in 1888 followed a leading figure of the day, David Cook, to Ventura County in California to participate in a Methodist community, the Second Garden of Eden. Cook was also from Elgin and had bought Rancho Temescal from the Del Valle family. He set about founding the town of Piru and building a Methodist Episcopal church.

Lenton moved to California in 1888 and stayed in Piru City until 1896, when he left for Los Angeles and ultimately Long Beach. During his career in California horticulture, Lenton introduced more than twenty-three new cultivars of carnation. He registered his new cultivars and often entered them in shows. Ventura was far from established centers of population, but in 1887 the Southern Pacific Railroad reached Piru City, and the U.S. Post Office arrived in 1888. Presumably, Lenton used these methods of shipping and transportation, but it was a long journey east, and one wonders how he kept his flowers fresh all that time.

He named some of his new flowers for people and places in California: 'Bidwell' commemorates John Bidwell, the first person to lead a party of emigrants from Ohio to California overland without losing a single soul, in 1841. The sources of the names 'Piru', 'Paradise', and 'Pacific' are obvious, while 'Ramona' celebrates a beloved fictional heroine. All these new flowers were introduced in the 1890s, even though his nursery remained open until his death in 1905.

His son, Albert, learned the gardening trade from his father and was briefly foreman of the groundkeepers at the University of California at Berkeley until he had to go back to Long Beach as his father's health declined.

Redondo Carnation Company, Redondo Beach, California

The company was named for the town in which it began: Redondo Beach, California. It appeared in a Redondo Beach directory in 1906, with Henry Feder listed as the manager. The directory also listed his home address. Between 1907 and 1910, Feder changed the name of his company to Carnation Gardens Nursery and moved it to another address. A local newspaper featured the company in about 1910: "Carnation Gardens . . . offered 12 acres of sweet smelling flowers almost always in bloom." Three years later, the nursery was part of a new subdivision, and by 1916, both it and Mr. Feder had disappeared completely.

Warren R. Shelmire (b. 1850), New Garden, Pennsylvania

Shelmire was born in Philadelphia in 1850. When he was nineteen, he moved out into the country and settled in New Garden, a very small town near Avondale in Chester County. In 1884, he started his own florist business in New Garden and began to experiment with carnations. He soon learned that this was a slow and frustrating business, yet he persisted until he introduced 'Eldorado', a really good cultivar. Shelmire was also known for 'Kitty Clever' and 'Eulalie'. He produced twenty-six cultivars.

As he continued to work with his seedlings, he found that the pollen of 'Golden Gate' almost always produced a yellow or yellow variegated flower. Shelmire pursued this line vigorously, making up to eighty new crosses a year at one time.

Charles Starr (1846–1891), Avondale, Pennsylvania

Starr came from an old Pennsylvania family and was the youngest of seven children. His earliest known forebear settled in New Garden Township in 1712. The Chester County history states that he was "one of the most successful florists in the State and especially noted as a carnation grower." Starr owned "Pleasantville Green-houses" in New Garden near Avondale on the Baltimore and Philadelphia Railroad. "He makes a specialty of carnation pinks, which, with his other innumerable varieties, he ships by rail to all parts of the country." The authors of the county history go on to say, "From small beginnings he has extended his green-houses and establishment to large proportions and does annually an extensive business." He produced forty-two carnation cultivars.

John Thorpe, New York

Thorpe produced twenty-two carnation cultivars (see chapter 2 for more information about Thorpe). Despite his renown and accomplishments, he was said to have ended his life in extreme poverty.

H. Weber and Sons

Henry Weber was born in Hessen-Cassel in Germany in 1835 but immigrated to the United States and settled in Oakland, Maryland. Weber had stayed in the standard German public school until he was fourteen, when he left to be apprenticed to a florist. He was so effective at his work that he was put in charge of all the greenhouses and gardens as foreman at a very young age. He joined the British Army in 1854 and fought in the Crimea; the army stationed him at many places around the globe during his ten years of service.

Weber's brother John had come to the United States before he did. The dates are not clear, but Henry probably joined him in 1865. For a time they were in business together, but Henry sold his share to his brother and opened his own business in Cumberland, Maryland. By 1879, he had bought another piece of land near Oakland and established his final enterprise. He was very

Richard Witterstaetter.
From Charles Willis Ward, The American Carnation *(1903).*
Reproduced compliments of Applewood Books, Carlisle, MA

active in the American Carnation Society and produced seven cultivars. Weber died in 1904 and was buried in his private cemetery. His six sons had gone into the business with him and continued the firm on their own.

Richard Witterstaetter (b. 1859), Cincinnati, Ohio

Witterstaetter was born in Sedamsville, Ohio, in 1859 and never really left the state. He began to breed carnations in 1890 and exercised extremely tight control over what he introduced. Ward says he introduced only four cultivars, 'Emma Wocher', light pink; 'Estelle', scarlet; 'Evelina', white; and 'Adonis', scarlet, but the register attributed another six cultivars to him as well. Because of his discipline and skill, he was elected to office in the American Carnation Society, and other breeders looked to him for guidance.

6

Clematis

CLEMATIS IS widely distributed across the temperate zones of the Northern and Southern Hemispheres. Experts believe there are more than two hundred species worldwide. Like so many other valuable ornamental plants, by far the largest number of species is found in China. *Clematis vitalba* L., is the only species native to the United Kingdom, and has long been used in gardens. Gerard called it "Traveller's Joy" in the 1597 edition of his *Herball*. The plant's beauty is due to the large colored sepals giving the appearance of petals.

Most clematis plants are true vines. True vines differ from herbaceous plants in that the stems have no intrinsic tensile strength but must use other plants or structures to support the weight of the plant above. The so-called climbing roses are not vines but ramblers. Their stems grow out so far and are so long that they need to seize on any support that will help them, whether it be vertical or horizontal. This characteristic has survival value. It enables the plants to propagate themselves.

During the sixteenth century, several species of clematis were imported into Britain from Europe. The wild species *Clematis viticella* L. came from Spain. Queen Elizabeth I's apothecary Hugh Morgan grew *C. viticella* in his garden before 1569. The flower came to be known as virgin's bower, possibly after Elizabeth, the Virgin Queen. It is one of the few plants that came from Spain in those early years. Traveling around Spain was difficult, and the people were not helpful, particularly to the English. Both *C. viticella* and *C. vitalba* have been useful as rootstocks.

Clematis flammula L. came into Britain at about the same time from the Mediterranean region. It had a very powerful fragrance but also strong purgative effects. *C. integrifolia* L. arrived from Hungary in 1573. The evergreen *C. cirrhosa* L. arrived in 1590 from the Balearic Islands off the coast of Spain, followed by *C. recta* L. from southern Europe in 1597. In all, five new species were introduced into Britain during the sixteenth century.

Like many other garden plants from that epoch, such as the sweet pea, nothing much changed for almost two centuries. People were content to grow what they knew. If a flower were exquisitely fragrant, like Father Francis Cupani's original sweet pea, gardeners were less concerned about its appearance.

Other kinds of clematis crept in subsequently. *Clematis florida* Thunb. came from China in 1776, and *C. alpina* (L.) Mill from mountainous regions of Europe in 1792. The double greenish white *C. florida* is the ancestor of many double cultivars. In the early nineteenth century, more species appeared in western Europe from a wider range of countries. *Clematis meyeniana* Walp. came from Hong Kong in 1820. Hong Kong was a barren windswept speck in the ocean in 1820. When Robert Fortune visited it in 1842, he said that it would never amount to anything. How little did he know!

Clematis campaniflora Brot. was found in Portugal and sent to Britain in 1820. The blue Himalayan *C. grata* arrived in 1831. Soon afterward, *C. heracleifolia* DC appeared from China. Mongolia was the source of *C. aethusifolia* Turcz. in 1850. *C. fusca* Turcz. came from Japan, as did *C. stans* Siebold & Zucc. The latter was useful for breeding, because it was a dwarf type.

All these species had small flowers. The excitement began when two species with very large flowers came from China and Japan. Phillip von Siebold took *Clematis patens* C. Morren & Decne with him from Japan to the Netherlands in 1836, shortly after Lady Amherst first planted the Himalayan *C. montana* Buch.-Ham. ex DC in England in 1831. Robert Fortune sent *C. lanuginosa* Lindl. & Paxton to England from China in 1850. It did not take long before orgies of hybridization started.

North America has its own native species. The sweet autumn clematis, *Clematis paniculata* J. F. Gmel. (syn. *C. terniflora*), is very vigorous and highly fragrant. The sweet autumn clematis has two other synonyms: *C. maximowicziana* and *C. dioscoreifolia* (obsolete). According to the Missouri Botanical Garden's Plant Finder, *C. paniculata* (syn. *C. indivisa*) is actually a separate species native to New Zealand. This plant arrived in Europe in 1840.

Other American species are *C. crispa* L. and *C. viorna* L. *C. crispa* crossed to Britain in 1726 at almost the same time as *C. orientalis* L. came from Asia in 1727. At the end of the eighteenth century, *C. virginiana* L. went to Britain from Virginia. The purplish-blue *C. verticillaris* DC was another American introduction in 1797. *C. addisonii* and *C. ochroleuca* are additional North American species used in modern breeding programs. The last quarter of the eighteenth century was one of wars and strife, yet new plants continued to be sent from one country to another. The miracle was that any of them reached their destinations.

The nineteenth-century imports ended with *Clematis songarica* Bunge from Siberia, *C. douglasii* Hook., and the Russian *C. tangutica* (Maxim.) Korsh. New species continued to be imported into Britain starting in 1900. Another Chinese species, *C. quinquefoliolata* Hutch., with its five-lobed leaves, was interesting enough, but the real star of that epoch was *C. armandii,* also found in China.

Clematis spooneri Rehder & E H Wilson (syn. *C. chrysocoma* var. *sericea*), *C. chrysocoma* Franch., and *C. macropetala* Ledeb. rounded out this cohort of species. In the next decade, the appearance of *C. fargesii* Franch. (now *C. potaninii* Maxim.) from China and *C. serratifolia* Rehder from Korea more or less completed the huge spate of imported species. The unusual *C. afoliata* with its rush-like stems came from New Zealand.

BREEDERS

ENGLAND

Thomas Cripps (1809–1888), Tunbridge Wells, Kent

Cripps was one of the earliest breeders of clematis. He opened his nursery in 1837, subsequently naming it Thomas Cripps and Son. He worked primarily with *Clematis patens, C. lanuginosa,* and *C.* 'Jackmanii' to produce his hybrids. A brief obituary of Cripps appeared in the *Gardener's Chronicle* of 1888 following his death earlier that year at the age of seventy-nine. The obituarist commented only that Cripps was known for introducing the first fuchsia ever to have white sepals in 1848 and left it at that. There was no mention of the forty-four new cultivars of clematis he introduced, many with the assistance of one of his sons.

The first few of these came out in 1866: 'Gloria Mundi', 'Guiding Star', and 'Lady Caroline Nevill'. In 1867, 'Victoria', 'Mme Van Houtte', 'Annie Wood', and 'Captivation' were introduced. The final one, 'Jackmanii Superba', appeared in 1880. George Jackman was very active in the 1860s, and perhaps Cripps was inspired to compete with and even to surpass his rival.

Cripps was quite well off and moved his premises more than once to obtain more land. Sadly, both of Cripps's sons died before he did, though it is not clear which one of them was the son referred to in the nursery's name. Thomas's daughter, Ellen, took over the business after her father died, and successfully maintained it. She never married. A woman running a large business was quite unusual for the period, but she was well respected in the town and given an elaborate funeral when she died in 1902. The nursery still managed to stay open after her death, and as late as 1907 the firm's catalogue listed seventy-five varieties of clematis. By 1912 it could no longer continue and had to close.

Jim Fisk (d. 2004), Westleton, Suffolk

Fisk died in 2004 at the age of ninety-two. His influence still remains, less through his own crossbreeding than through his introductions of significant European cultivars. Fisk was born in Suffolk and at the age of fourteen started his career as a gardener at Notcutts Nursery in Woodbridge. After serving in the Royal Navy during World War II, Fisk used his demobilization money to set up his own nursery specializing in clematis at Westleton, also in Suffolk. The nursery ran for almost fifty years. He finally closed it in 1999 at the age of eighty-seven.

Jim Fisk had very little time during this busy life to acquire much in the way of formal education, but he wrote several very useful books about clematis, which are still read. Very late in his life he finally received some recognition of his achievements with the award of an MBE (Member of the British Empire) from the Queen in 1997. One of his most enduring contributions was to introduce the work of Brother Stefan Franczak, the monk in charge of the garden at the Jesuit monastery in Warsaw, at a time when Eastern Europe was a separate world.

Barry Fretwell, Peveril Nursery

Fretwell is still alive but has recently retired. He played a major role in the modern development of clematis in England. Magnus Johnson, a Swedish plantsman and creator of a reliable catalogue, only listed a few of his cultivars, but *Clematis* 'James Mason' is an outstanding one, in the purest of whites.

George Jackman (1801–1869), Woking, Surrey

In England, George Jackman of Woking, Surrey, led the way. Jackman was not the first recorded nurseryman to cross clematis, but he did things on a very large scale.

George Jackman, known as George I in the family, was the grandson of the founder of the nursery in Woking, William Jackman. He and his brother Henry inherited it in 1830 and ran it together for about ten years, but it did not prosper, and they parted ways. George continued to be in charge and later developed the nursery into a very successful business, eventually employing thirty-five men.

George's son George II (1837–1887) took it to the next level. While he focused on clematis, he was too shrewd a businessman to allow one plant to dictate the fate of his firm, and Jackmans offered roses, other shrubs, and many trees. George II also capitalized on the value of the nursery's land, selling it for development and moving the nursery to another site. (The economist Maynard Keynes's great-grandfather, the dahlia "king" in Wiltshire, did something very similar, investing in property all the time he ran the nursery. As a result, Maynard Keynes was born to considerable wealth.) The Jackman nursery lasted as a family firm until 1967. It had a run of more than one hundred years.

In 1835, a Mr. Henderson of Pineapple Nursery, St. John's Wood, London, introduced the first known intentional cultivar, 'Hendersonii' (now known as *C.* × *eriostemon*), derived from a cross between *C. viticella* and *C. integrifolia*. In 1858, the Jackmans decided to work on clematis systematically.

Clematis 'Hendersonii'.
Shutterstock.com

They released *Clematis* 'Jackmanii', the result of crossing *C. lanuginosa* with 'Hendersonii'. They had selected this cultivar after examining three hundred seedlings. In the ensuing years, Jackmans released dozens of cultivars, often named for the prominent customers who patronized their firm. Very few of these cultivars survive today, but *C.* 'Jackmanii' is still grown in some gardens in England.

When George Jackman died in 1887, Arthur George Jackman took over the nursery. Arthur's brother Percy joined him a few years later. They weathered the disasters of the First World War but Arthur died in 1926. His son, George Rowland Jackman, always known as "Rowland," became the nursery manager. Once Percy died, Rowland was the sole proprietor, keeping the firm going until 1967. He was not enthusiastic about clematis but did introduce three good cultivars still grown today. Rowland Jackman died in 1976.

Ernest Markham (1880–1937)

Markham took care of William Robinson's gardens at Gravetye Manor, a position one might imagine to be very difficult to hold because Robinson was notoriously irascible, yet the two men seemed to have got on well enough together. Robinson had been seriously injured in an accident in 1909 and depended heavily on Markham.

Markham distilled his rich knowledge of the genus into a useful small book, and Robinson wrote a very generous foreword for it. So much for the standard myths of Robinson's choleric personality. That may have been a persona he adopted to get his points across. This was Robinson's last piece of writing. He died in 1935 at the age of eighty-eight.

Markham had taken charge of the clematis collection Robinson was said to have bought from the Veitch nurseries in 1922. Sir Harry Veitch sold off all his inventory in 1914, as there was no suitable heir to live up to his high standards. There was, however, a Veitch nursery in Exeter, owned and operated by a remote cousin and still open in the early 1920s. Possibly this was where Robinson found his collection.

Markham left ten known cultivars, including one with the charming name of *Clematis* 'Stolen Kiss'. He named one cultivar for his wife, 'Mrs. Markham', and another 'Gravetye Manor'. This one had fine dark red sepals.

He was an important figure in the local village, as he enjoyed playing cricket and was very keen on the game.

Ernest Markham helping visitors at Gravetye Manor as a young man.
Reproduced by permission of Kenneth Woolfenden

Ernest Markham on his wedding day.
Reproduced by permission of Kenneth Woolfenden

Ernest Markham with his employer, William Robinson.
Reproduced by permission of Kenneth Woolfenden

Charles Noble (b. 1817), Sunningdale, Berkshire

Noble was born in 1817 but the date of his death is not known. What is known is that he retired in late 1896 and sold all his plant stock. He also insisted that all his correspondence be burned, a huge loss for history, as he received letters from the great collectors like Nathaniel Wallich.

Noble's manner belied his name. There was nothing noble about his behavior. He was said to be ill-tempered and unpleasantly mean, in both senses of the word. This unusually forthright assessment was recorded by Eleanor Willson in her useful small book *Nurserymen to the World,* a detailed record of garden businesses in Surrey.

Noble started his career by joining John Standish's (1814–1875) prosperous nursery on Bagshot Bridge, not too far from Windsor, in 1846. A year later, they opened new premises a few miles away, the Sunningdale Nursery, but still retained the old place as growing grounds. John Standish was a man of large imagination and vision.

He was also a fine hybridizer. One of his outstanding achievements was to obtain some of the earliest plants to come from China, soon after Robert Fortune returned to England in the early 1840s with his spoils. Standish trumpeted his success in his catalogues. The new plants elicited awe and amazement, or perhaps "shock and awe."

Standish and Noble's partnership lasted for eleven years but then broke up. John Standish put it this way: "There was no room for two suns in Sunningdale." For a time they were not on speaking terms, but later they patched up their differences. Noble kept the Sunningdale nursery, while Standish moved to Ascot and opened the Royal Nursery.

Eleanor Willson does not mention Sunningdale Nursery's clematis, but she states that Standish did all the plant crossings. Noble seems to have been more interested in the business side of things. In spite of that assumption, sixty cultivars of large flowered clematis are attributed to Noble in Johnson's list. More than half of them were introduced after the two partners split up. It is possible that Noble employed a plant breeder but did not release the person's name. Only an additional six cultivars appear solely under Standish's name.

Walter Everitt Pennell (1910–1977), Lincolnshire

Pennells has a long and illustrious history in the English nursery business. Richard Pennell started his nursery in Lincolnshire in 1780, just a few years after Veitch, but unlike Veitch, his family still has a firm to run. From the information on the Pennell website it is clear that they have understood the secret of modern business. Continually modifying and adapting to change is the only way to thrive. Staying with the same strategies is actually going backwards.

Walter Everitt Pennell was a lineal descendant of Richard Pennell. Clematis had become important in the mid-nineteenth century, and Pennell's catalogue for 1846 listed seven species of clematis. By 1880, many more clematis species, as well as new cultivars, began appearing. In 1939, the catalogue listed forty-eight large-flowered cultivars and twenty-nine species of clematis.

As such businesses grow and prosper, the next generation can be better educated in a way that benefits the firm. It was even possible now for some of the children in a large family to refuse to work in the family business if they wanted to do something else.

Uno Kivistik, surrounded by his clematis.
Reproduced by permission of Kenneth Woolfenden

Walter Everitt read chemistry at Reading University, and really did not want to become a nurseryman, but in the event he rather reluctantly took on his responsibilities. His career was interrupted by the Second World War, but once he had completed his service in the Royal Artillery he returned to work.

His scientific background led him to look at crossbreeding as a satisfying occupation, and he began to introduce many new and effective cultivars of clematis. The best-known one is named for his wife, *C.* 'Vyvyan Pennell'.

ESTONIA

Uno Kivistik (1932–1998)

Uno Kivistik worked with his wife, Aili, throughout his career. Together they introduced some 150 cultivars of clematis, more than anyone else. These were selected from the more than 6,000 hybrids they bred. This was all accomplished between 1979 and 1996, with about eight or nine cultivars produced per year. Mrs. Kivistik is still alive and working at the nursery.

Estonia, which is in northern Europe, has the typical punishing winters and short summers of that part of the world. Part of the Kivistiks' goal was to find or create cultivars that could survive and even flourish in these conditions. Most of the plants available to them were Estonian, with very few foreign choices. Estonia was in the Communist bloc.

Kivistik was one of four brothers whose father had been a plant breeder. They worked on fruit such as apples, pears, and grapes, and also roses to suit the climate. Uno Kivistik collected clematis from all parts of the Baltic region and the Soviet Union. He founded the Estonian Clematis Club and was the chairman for many years. Since his death, his son Taavi and daughter-in-law Aime have been carrying on the family tradition. Two cultivars were named for Uno Kivistik: one in the USSR by M. Sharanova in the 1970s, and one by his children, *Clematis* 'Uno Kivistik', more recently.

FRANCE

Auguste Boisselot, Nantes

Boisselot was a nurseryman in Nantes. He probably died in 1870, although the records are not clear. In addition to his other work, Boisselot was a skillful orchardist and successfully grafted vines. In 1859 his work on clematis was mentioned in an article in *Revue horticole*. Little else is known about him.

Alfred Carré (1841–1912), St. Julien-les-Troyes

Carré lived and worked in St. Julien-les-Troyes. He is best known for the cultivar *Clematis* 'Gloire de St Julien'. The town stationer employed him as a gardener, and in 1887 he won a silver medal at one of the Société d'Horticulture de l'Aube's shows for his work on roses and carnations. Twelve clematis cultivars are attributed to him. The president of the Société d'Horticulture de l'Aube at the time was Charles Baltet, a distinguished flower breeder himself. From 1880 until his death, Carré was the administrator of the Société Horticole Vigneronne et Forestière de l'Aube. Carré's father had been a notable nurseryman with an interest in clematis.

Louis Christen (b. 1828), Versailles

Christen was a nurseryman in Versailles. Magnus Johnson listed fifteen new cultivars bred by him. In 1879, Christen published *An Illustrated Encyclopedia of Clematis*. Christen won the silver medal at a show in Paris in 1892 for displaying twenty-five varieties of clematis. He also entered shows in other parts of the country, such as the one held by Lyon Horticole in Lyon, where he displayed *Clematis* 'Papa Christen', 'Madame Boselli', and 'Eugene Delattre'.

In addition to these accomplishments, Christen was also a municipal councilman. The census records indicate that his nephew Charles Rink lived with the family and worked as a nurseryman too, perhaps with his uncle. There is no record of when Louis Christen died.

Victor Lemoine (1823–1911), Nancy

Clematis was only one of the genera to which Lemoine made significant contributions (see earlier chapters and my previous book). More than fifty cultivars by Victor and, later, his son Emile Lemoine are listed in Magnus Johnson's book.

Francisque Morel (1849–1925), Lyon

Morel was one of the amazing coterie of distinguished nineteenth-century nurserymen in Lyon. In his lifetime he was noted as a landscape architect, but clematis was a significant part of his work. Francisque Morel had a special way with water features in parks and a profound knowledge of the ways in which trees could be used in the landscape.

Magnus Johnson attributes nineteen clematis cultivars to him. Later in life, Morel's son joined him in their business. Morel published many articles describing his new cultivars in *Revue horticole* and *Lyon horticole* starting in 1884. Morel introduced *Clematis* 'Ville de Lyon' in 1899 and 'Comtesse de Bouchard' in 1900.

Clematis 'Ville de Lyon' was bred from *C.* 'Madame Edouard Andre', introduced by a Monsieur Baron-Veillard in Orléans in 1893. The seedling that became 'Comtesse de Bouchard' was given to the countess as a special honor. Its parents were *C.* 'Viviand Morel' and *C. coccinea*. Another exciting hybrid

Clematis 'Ville de Lyon'.
Photograph by Doug Wertman, Shutterstock.com

Morel bred was *C. pitcheri* crossed with *C. coccinea*. The violet *C. pitcheri* came from Texas, while *C. coccinea* imparted a deep carmine color.

A man with a strong sense of civic responsibility, Morel also served as a vice president of Lyon Horticole. It is a commentary on his reputation and accomplishments that numerous representatives of horticultural organizations attended his funeral in 1925. The president of the Société Lyonnaise d'Horticulture, Professor Gérard, gave the eulogy. He called Morel an artist in both landscape design and flower breeding. Morel had a unique combination of natural talent, education, and the rich background gained from watching his father, François, work as an arborist. François was part of the distinguished company of pomologists and arborists in and around Lyon at the time.

Gérard pointed out that Morel understood the value of botany to his work and read very widely in other languages. His colleague, Joseph Victor Viviand-Morel, was equally erudite, and together they formed a formidable team. The latter was also at one time president of the Société Lyonnaise d'Horticulture. Even allowing for the inevitable hyperbole of a eulogy, one can see that Morel was a truly distinguished man and deserves to be rescued from the oblivion into which he has fallen ninety years after his death, becoming a mere name on a page.

Moser

Johnson lists eight cultivars for Monsieur Moser, including 'Nellie Moser' and 'Marcel Moser', most of them in the 1890s and one in the following decade, before 1910.

JAPAN

Kazushige Ozawa (d. 2003)

Ozawa's *Clematis* 'Asao' and 'Kakio' are perhaps the best known of his work. He used the small-flowered North American species for a very specific purpose, adding *C. integrifolia* to increase the size of the individual flowers. Practitioners of the Japanese tea ceremony like to have a single clematis flower in a vase on the table to enhance the ritual, and Ozawa set out to breed long-stemmed, rich yet subtly hued flowers with interesting shapes for that purpose. This led him to run a successful cut flower business. The American clematis breeder and expert Maurice Horn got to know Ozawa when the latter came to the United States twice to visit the breeders from whom he had obtained the seed.

POLAND

Brother Stefan Franczak (1917–2009)

Brother Stefan was twelfth in a family of thirteen children. His father was a farmer, and as a young man Stefan attended the local agricultural college. He later taught at another agricultural college. After World War II ended, he changed course, and in 1948 he entered the Jesuit monastery in Warsaw. He decided he did not want to become a Jesuit priest but chose to be a monk instead. Because he had an agricultural background, the authorities put him in charge of the monastery's extensive kitchen gardens.

Franczak had the misfortune to join a religious establishment at the time when the communists took over in Poland. From being part of the dominant governing structure of the country, the Catholic Church became a pariah. For example, when the priests wanted to build a new church on their property, the government refused to let them do it, because the communist government objected to the monastery owning any private land.

To drive the point home, the government then condemned the monastery's land, leaving them only a fraction of a hectare, but they reckoned without Jesuit intellectual power and resourcefulness. To counter the argument of selfish elitism, the priests turned their vegetable plot into an open public garden for everyone to enjoy.

I had wondered why and how Brother Stefan decided to grow clematis. In a recent memorial essay about Brother Stefan, Ken Woolfenden pointed out that in the 1950s the old kitchen garden grounds were full of fences and old tree stumps and could not easily be turned into flower beds. Climbing vines were the answer.

Brother Stefan Franczak with some of his clematis in Warsaw.
Reproduced by permission of Kenneth Woolfenden.

Brother Stefan started out by buying a few plants and then propagating them himself. As his collection grew, he sold the extra seedlings, thereby raising more money to buy additional clematis plants. He was also able to tap into the huge European Jesuit fraternity and gain more plants that way too. Spontaneous hybrids and volunteer plants intrigued him, and he soon began to make deliberate crosses.

The garden eventually contained more than 900 varieties of ornamental plants, mainly clematis but also some hemerocallis and iris. Warsaw's citizens took great pleasure in their garden. Once Cardinal Karol Wojtyla was elevated to the papacy in 1979, the communist government eased its attitudes. Shortly after Wojtyla became Pope John Paul II, the Jesuits were given permission to build, and the garden was reduced to a mere shadow of its former self. Today, Woolfenden reports that there is nothing left. All the land is now taken up with new buildings.

The Franczak cultivars are noted for their brilliant color, sturdiness, and resistance to disease. Brother Stefan was particularly partial to bright colors, stiff sepals, and contrasting stamens. The English clematis expert Jim Fisk first came across the Franczak cultivars in the 1970s, and they worked together. In 1982, Fisk showed *Clematis* 'John Paul II' at the Chelsea Flower Show. Brother Stefan was allowed to go to London for this event and even appeared on English television.

As Brother Stefan grew old and sick, he was transferred to the monastery at Gdynia to spend his remaining days. He left a legacy of eighty-two cultivars, many of them named for major political and religious figures of his time. Perhaps the best known is *C*. 'John Paul II'. Many are no longer to be found, victims of time and neglect once Brother Stefan ceased to be active. In 2009, the year of Brother Stefan's death, the Polish prime minister Lech Kaczynski personally presented him with the government's highest honor.

Wladislaw August Noll (1900–1979), Warsaw

Noll took part in the doomed Warsaw Uprising of 1944 as an infantry captain and miraculously survived not only the fighting but also the slaughter by the Germans and subsequent witch hunt by the communist authorities. In 2002, his grandson Andrejz Frycz received long-overdue medals awarded posthumously for Noll's bravery and heroism in defending the homeland.

In a recent memorial essay, Szczepan Marczyński provides a picture of Noll's life together with a good deal of previously unknown information. Noll was not a nurseryman but had several different occupations as a young man. The fateful encounter with clematis occurred in 1928. After completing his agricultural studies, he was engaged for a time in growing and collecting medicinal plants, and met the Weiss family in Zloczow during that period. They had a collection of clematis and gave him four plants. That was all it took. He was smitten. For the rest of his life, Noll grew clematis wherever he lived.

Once he settled in Warsaw after the war, he devoted himself to his plants. In 1969, he bought a larger piece of land about twenty kilometers south of the city and was able to expand even more. At its peak, his collection had more than two hundred cultivars. Noll raised and selected his own varieties and named them for prominent people and events in Poland. Space was still an issue, and he sometimes commissioned other people to propagate his cultivars, including Brother Stefan Franczak. Noll was extremely interested in

Clematis 'Niobe'.
Photograph by Dave Denby, Shutterstock.com

everything that had to do with clematis and wrote to many experts all over Europe. Marczyński says the correspondence with Jim Fisk was both lively and voluminous.

About twenty years ago, there was an unfortunate misunderstanding over a cultivar bred by Brother Stefan, which some people accused Noll of expropriating and renaming. Noll's family considered the author of the memorial essay, Szczepan Marczyński, to be the source of this accusation, and they were initially quite hostile to him. Brother Stefan himself believed that Noll's cultivar 'General Sikorski' was identical to his own 'Jadwiga Teresa'. He thought this was because he had previously given Noll some unnamed cultivars. In the end, Marczyński examined Brother Stefan's notebooks and showed that the accusation was not true.

The world is the richer for Noll's 'Niobe', a luscious red climber, and 'General Sikorski', with mid-blue flowers and also a climber. Both cultivars may be grown in containers. Another of his glorious cultivars is the one named for his daughter Halina. It has both single and double white flowers, depending on whether they emerge on new or old wood.

Clematis 'Nikita'.
Photograph by Sergei Afanasev, Shutterstock.com

RUSSIA (USSR)

Margarita Alexeevna Beskaravajnaja (1928–2003)

Beskaravajnaja was born in a small town in southwestern Russia and attended the university at Voronezh. She must have been a very able student to be allowed to enroll at the Moscow State University for postgraduate studies in botany and plant science. This was in the late 1940s, when the USSR was still profoundly depleted by the Second World War. More than twenty million men had died, and perhaps this was how a woman received a place. As with the case of Mikhail Orlov (see below), she must have been completely apolitical.

Margarita Alexeevna Beskaravajnaja.

After university, she took a job at the State Nikitsky Botanic Garden in Yalta and worked there for more than thirty years. A. N. Volosenko-Valenis joined the staff while she was there. After he died, she took over his unique collection of clematis and continued his work with the assistance of Helena Donyushkina. Although Beskaravajnaja was far from the center of things, she kept in close contact with the rest of the horticultural world and wrote books and articles about her work. She introduced more than two dozen cultivars, among them *Clematis* 'Aliosha', familiar to many Western gardeners.

Mikhail Orlov (1918–2000), Kiev

Orlov was educated at the Forest-Technical Academy in Leningrad and later received a doctorate from that institution. The 1940s were a turbulent period in USSR agriculture, but presumably Orlov kept to his studies and did not express any political views. (This was the time when Trofim Lysenko made his moves and destroyed the career of the botanist and geneticist Nikolai Vavilov. Vavilov was in charge of the Soviet agronomy program and a widely respected scientist.)

Orlov transferred to Kiev (now Kyiv) in Ukraine and ran a clematis breeding program there, trying to find cultivars that would resist wilt, the fungus which was a death knell for many of these plants. He introduced about forty plants, but they are not known in the West.

Trofim Lysenko was a scientific opportunist. Food was very scarce in the Soviet Union at the time and Stalin hoped to be able to grow grain above the usual northern latitudes. Lysenko was a follower of Lamarck and thought that growing plants in the cold section of the country would transform their genes. When Vavilov told Stalin it would take ten years to do this and Lysenko said he could do it in three, Stalin chose Lysenko. Not content with this triumph, Lysenko made sure Vavilov was completely destroyed. Vavilov starved to death in a gulag.

Maria Fodorovna Sharonova (1885–1987), Moscow

Sharonova was a remarkable woman. The fact that she lived to be 102 years old through a tumultuous century in which the Russia she had known became the USSR and an exceedingly dangerous place to live was amazing enough, but what she did with the extra valuable years she survived deserves even more respect—no "golden years" in a rocking chair for her.

After college, she became a botanist and worked for many years on dahlias. Very late in life, Sharonova decided to breed short clematis plants in bright colors, suitable for Moscow balconies and window boxes. What inspired her to do this is unclear, but even in her late nineties and bedridden she was still active in this program, though it was being implemented by other people.

Moscow's climate is forbidding with very long winters and short summers. It is so far north that there is not enough time for clematis to set seed every summer. Sharonova might not get seed for two or even three years. Eventually, she left more than two hundred cultivars, of which forty were named and introduced. Although she set out to create shorter forms, there were also taller, very free-flowering ones among this number. The names are all in Russian, sadly few of them familiar in the Western world.

Alexander Nicolaevich Volosenko-Valenis (1928–1967), Yalta

Volosenko-Valenis was born in Belarus. In 1951, he took up a position at the State Nikitsky Botanic Garden in Yalta and began his clematis breeding program. By 1967 he had developed many of the first Russian cultivars and published nine articles about them. Volosenko-Valenis died very young, at the age of thirty-nine. Had he lived, he might have gone on to even greater achievements, but as it was M. A. Beskaravajnaja, a skilled woman flower breeder, took over where he left off.

SWEDEN

Magnus Johnson (1907–2002)

Johnson became a gardener in the early 1930s and very quickly found his calling in clematis. As soon as he could, he started his own nursery, and ultimately introduced more than 120 new cultivars of clematis. Johnson

Magnus Johnson.
Reproduced by permission of the Magnus Johnson family

sought out many plants from other countries and developed a methodical breeding program.

In 1952, the City of Gothenburg appointed him head gardener at the town's botanical garden in 1952, a position he held for eight years. The Royal Horticultural Society awarded him its Veitch Medal in 2002, shortly before he died. That was a signal honor, not often given to non-English people. His monograph on the genus clematis has never been surpassed. One of its few weaknesses, according to Linda Beutler, is the absence of information about Japanese breeders. Those developments were taking place as he grew older, and evidently he was not aware of them.

7

Pansy / Viola

WILD VIOLETS have been beloved by all for centuries and frequently celebrated in poetry and song. Wild pansies were familiar to country people who lived near woods and called them "heartsease." Approximately five hundred species of the genus *Viola* can be found in temperate zones worldwide. *Viola* is highly pleomorphic, and the fragrant and herbaceous forms are of particular interest in this context. Many are annuals, but in the right climate they may be perennial. Cultivated violas fall in the *Melanium* section of the genus.

The modern cultivated pansy can be said to have come into being in the early years of the nineteenth century. Although not morphologically close to any wild form, it belongs to the same genus. When *Viola tricolor* was crossed with other species of viola, the resulting hybrids were eventually known as *Viola* × *wittrockiana*. The name honored Veit Brecher Wittrock, the fourth Bergianus Professor of Botany at Stockholm University and founder of the Bergianus Botanic Garden. Wittrock had studied variation in the viola and published several papers on the subject.

Viola × *wittrockiana*.
Photograph by belizar, Shutterstock.com

Wittrock traced out the elements that went into making the modern pansy. *Viola tricolor* L. was the most significant ancestor, *Viola lutea* Huds. was another important part of it, and *Viola altaica* Pall. is a third, albeit rather minor. Given this complicated ancestry, Wittrock felt unable to come up with a suitable name for the modern version. He settled for calling them *Viola* × *hortenses grandiflora*. This has since been replaced by the current *Viola* × *wittrockiana*.

Some of the distinguishing features of the modern pansy are the black or dark-colored blotches, which are not found in the species flowers. These markings are caused by thickening and consolidation of the rays. The rays are nectar guides for insects. In plants grown from seed, care has to be taken that the blotch be opaque, but if the crop has not been carefully rogued, the original rays can still be seen when some flowers are held up to the light. ("Roguing" is the removal of any nonconforming plant from a commercial crop grown for seed. If rogues' seeds are inadvertently included in the final product, customers feel cheated. Roguing is very labor intensive and hence expensive.) The size and configuration of the blossoms are also different, with larger and flatter petals held horizontally, presenting a "face."

In general, gardeners did not use the existing pansies for bedding, because they were straggling and untidy in habit. Instead of trying to overcome these defects, breeders primarily focused on the range of color and size of the blossom. Only later, starting in 1869, did one of Dickson's employees in Edinburgh, James Grieve, undertake to improve the growth and habit of pansies. He turned the slightly unreliable flower into a useful bedding plant (see the discussion of James Grieve below).

In the nineteenth century, the amateur florists strove mightily to create a pansy with a completely round "face." The petals had to overlap in such a way that none of the natural waves or curves obtruded. These came to be known as "show pansies."

The florists had regular meets and competitions and developed very idiosyncratic ways of presenting their treasures. In many cases, the pansies were laid flat on a sheet of paper that would then act as a collar. This was so unnatural that Gertrude Jekyll, the famous Victorian gardener, protested and asked for a better way of doing things. Although she was "gentry," she knew about the florists' societies and their activities. She argued that much of the beauty of the flower lay in its gentle and attractive curves. Just as she had lamented the bad reputation of the marigold, she lamented the way in which the florists distorted a lovely plant.

The founders of the modern pansy began their work in the first quarter of the nineteenth century. Within about twenty-five years they had altered the humble little woodland flower beyond all recognition. Some of the first steps were taken in Buckinghamshire. In 1813, Admiral Lord Gambier presented his gardener, William Thomson, with various pansies he had collected in the wild around Iver, his Buckinghamshire estate, and suggested that Thomson cultivate them. Thomson slowly began improving the pansies, crossing them with several cultivated varieties, including a completely blue flower from a Mr. Brown's nursery at Slough. At one point, he probably also used a yellow Siberian viola, *V. altaica*.

Thomson was very proud of his initial successes but looking back many years later recognized that they paled in comparison to the ones derived from a volunteer plant growing among his heathers. He recalled that he was startled to see what he thought was a tiny cat's face staring up at him from the bed. This was the first time the characteristic modern blotch was noticed. Thomson began working with this sport and introduced 'Madora', followed by 'Victoria' and numerous descendants of that line. The actual date of his discovery is not clear, but by 1841 he had written a retrospective article about his work in the *Flower Gardener's Library*, reprinted in the *Floricultural Cabinet*.

At the same time as Thomson was working for Lord Gambier, Lady Mary Elizabeth Bennet and her gardener, William Richardson, were working on a similar project not far away. Lady Mary was the youngest child of the Fourth Earl of Tankerville, Charles Bennet, and she lived at the family home noted for its gardens, Mount Felix in Walton-on-Thames, Surrey.

The prominent Hammersmith nurseryman James Lee visited the estate early in this process, and saw a more attractive pansy, or "heartsease," as he called it, than the usual form. He encouraged Richardson to persist, and within a fairly short time Richardson introduced about twenty new cultivars. Richardson and Thomson were working about ten miles apart, and it is highly probable that they knew each other even though travel in those days was slow and cumbersome.

In 1839, Charles M'Intosh, the King of the Belgians' gardener, wrote *The Flower Garden*. He commented on the fact that James Lee in Hammersmith had obtained an all-blue flower from Holland, probably the same kind that Richardson had used at Walton years earlier. M'Intosh listed one of Thomson's cultivars, 'Lady Gambier', a large yellow flower, all those years later because it remained a successful plant. The blossoms were getting larger with each introduction, though there was still very little variation in color. M'Intosh was a very assiduous writer, publishing numerous useful horticultural and agricultural books for at least another thirty years.

English pansies reached France and Belgium in about 1840. The French and Belgian breeders, who were not bound by the stultifying rules of the English florists, worked on them and introduced larger flowers with more natural-looking shapes. The resulting plants began to reach England in about 1850. Once there, they were further modified but were still known as Belgian pansies at the time. William Dean, one of the expert Dean brothers, first referred to them as "fancy," a term still in use today.

Fancy show pansies by William Dean.
Reproduced by permission of Elizabeth Farrar, Pansies, Violas and Sweet Violets *(1989)*

The English nurseryman John Salter (see chapter 2) returned to London from France and started the Versailles Nursery in Hammersmith. When he left France, Salter took seeds of the new pansies with him back to London and offered them in his 1851 catalogue.

Salter introduced some very dramatically different types of pansy. His experiments in France had yielded striped and "stained" blossoms. He took those seeds to London and propagated the results. Other floral experts scoffed at these flowers, calling them "French rubbish," but the public liked them and the seeds sold well. This is all recorded in Wittrock's useful compendium. Wittrock describes one particular flower at a Royal Horticultural Society show in 1851 that had a mixture of streaky and flamed petals and no actual blotch. Instead, the flower had "branched honey stripes of an unusual colour, namely red."

Salter was successful in London for another few years, but by 1859 he had to cease growing pansies. Evidently a pest of some sort emerged that made the cultivation of pansies impossible. John Downie later became the most prominent of the men growing these new pansies, but it was Andrew Henderson who actually introduced the new forms to English gardeners on a large scale in 1858. He turned to William Dean to propagate them in large quantities at Dean's property in Shipley, Yorkshire.

The success of the Dean plants showed that the slightly cooler temperatures in northern England favored their growth. Gradually, pansy production established itself in Scotland for the same reason. William Paul, the firm of Downie, Laird, and Laing, and the Dickson Nursery (which later became known for the breeding of roses) all helped to build this Scottish industry.

John Downie did well with pansies he imported directly from Auguste Miellez near Lille. William Dean dutifully worked with Miellez's flowers, but then created his own line of pansies he had bred himself. Among these were 'Princess Alice' and 'Etoile du Nord', from 1861.

Although the pansy was bright and pretty, as noted before it did not do too well as a bedding plant. It was tall and a bit straggly, and the flowers were not reliably carried above the foliage. The Scottish gardener James Grieve (1841–1924) changed all that. The stimulus is said to have been the elegant displays of plant fancier John Fleming's crosses of *Viola odorata* with the new Belgian pansies in the early 1850s. If indeed Grieve did see this show, he

James Grieve, as portrayed in a watercolor by Scottish artist Henry Wright Kerr. *Credit: Henry Wright Kerr. James Grieve, 1841–1924. Horticulturist. National Galleries of Scotland.*

would have been a very young apprentice at the time. He began his career at the age of twelve and only moved to Edinburgh as an adult.

With all this activity, it is not surprising that pansy societies began cropping up in the middle of the nineteenth century. The first one, Hammersmith Heartsease Society, did not last very long. The Scottish Pansy Society was founded at the same time but did well and lasted for much longer, holding many memorable shows. Offshoots followed: the West of Scotland Pansy Society, the West of England Pansy Society in Exeter, and the Midland Counties Pansy Society in Birmingham are just a few of them. Finally, in 1898, the London Pansy and Viola Society was founded.

Development continued in France toward the end of the nineteenth century. Jules Bugnot in St. Brieuc, Brittany, and Cassier, James Odier, and Alexandre Trimardeau in Paris led in these exciting changes. Contemporary sources also indicate that a lot of work was being done in Germany at the same time. One well-known English nursery, Vaughans, informed its clientele that their pansy seeds came from both French and German growers, starting from 1889 right up to the beginning of the First World War. Vaughans's 1916 list was still very impressive. They offered more than fifty foreign cultivars. "Imperial German Pansies" were sold by Conard and Jones in the United States in 1900.

Several well-established German nurseries carried extensive lines of pansies, some bred in-house and others imported. Ernst Benary founded one of those nurseries. The firm of Benary celebrated its fiftieth anniversary in 1893 and at the time offered more than forty varieties of pansy. The list contains individual entries for seeds from Bugnot, Cassier, and Trimardeau. Cultivars named for Kaiser Wilhelm, Lord Beaconsfield (the former Mr. Benjamin Disraeli before being elevated to the peerage), and Victoria (presumably Queen Victoria or her daughter the Kaiserin) were also listed.

At present, Benary still offers many pansies. The name of one of their firm's most popular lines is Cats, a whimsical reference to Thompson's original observation, yet these F1 hybrids have rays or "whiskers" rather than blotches on the petals. Benary offers at least two other lines of F1 hybrids: Fancy and Joker. All the series have a wide range of very brilliant colors. Even during World War II, Benary introduced some new pansies. Presumably the ones that came out in 1941 had been in the pipeline for some years before, as practically none were introduced during the next eight to ten years. During the war, the firm switched all its effort to growing food for the German army.

Work on the violetta continued for many years. In the 1920s, D. B. Crane introduced three attractive new cultivars: the pale yellow 'Diana', the pale blue 'Eileen' with its golden eye, and 'Winifred Philips' in blue and white.

Much of the information recounted above comes from the work of Roy Genders. Genders, a slightly improbable English plantsman whose range of interests was extraordinary, produced a solidly useful book about pansies and violas in 1958. It seems that he no sooner saw a flower than he wrote a book about it and a very effective one at that. One of his contributions, a dependable feature of all his books, was listing the best cultivars for the reader to grow.

The British Library lists 160 separate titles, essentially a complete set of his work. Many are duplicates, but there were still at least a hundred individual books on flowers and other topics such as raising mushrooms and training greyhounds for racing. Genders was one of the first garden people to have a show on English television. In 1949, he was the "Mushroom Man."

This was all done when finding information was hard, requiring a lot of time and effort. The books were all written in longhand, not typed. For most people, this would be a full-time job, but his children, with whom I have had the privilege of speaking, all say he was a fun-loving father who spent a lot of the day in the garden or doing construction on his property. None of them has any idea how he found out everything he used in his books.

Genders recounted that both large firms like Suttons Seeds, and small businesses such as that of Carl Gustav Engelmann in Saffron Walden, offered many versions of the fancy pansy. Engelmann, known for his work on carnations (see chapter 5), introduced the pansy 'Engelmann's Giant' in the 1920s, although he did not give the flower that name. That was bestowed upon it by a gardener in a London park. Engelmann had generously distributed his excess seed and given some to the superintendent of one of the London parks. When the pansies were displayed, they made a very brave show. Passersby kept asking what they were called, and in desperation the gardener told them, Engelmann's Giants.

Engelmann was not alone. Carter, Dobbies, King, Hurst, Harrison, and Read, some of whom are mentioned in my earlier book, all introduced their version of large-flowered pansies in the 1920s. The Germans and French also developed beautiful large-flowered pansies, such as the Masterpiece strain in Germany. Considering the terrible conditions in Germany after the First World War, it is all the more impressive that some gardeners were still able to work on breeding flowers.

The Roggli family in the Alps near Lake Thun in Switzerland introduced Swiss Giants. Many of these varieties did better in the cooler north of England. When grown in the relatively warmer south, they blossomed very well for a short time but then petered out. In the north they bloomed for longer periods, giving much better value for park authorities with tight budgets. There were at one time about fifteen cultivars of Swiss Giant.

The pansies developed by Trimardeau, a French enthusiast, continued to be very popular even into the middle of the twentieth century. In 1958, Genders commented that the seeds sold in the millions, and indeed they are still available in the twenty-first century. The public likes them because the plants blossom freely and remain in bloom for a long time. There were half a dozen cultivars in the 1950s.

The tufted pansies or bedding violas are also very popular, especially because they are perennial. Genders ran some trials himself and quoted William Cuthbertson at Dobbies Nursery in Edinburgh doing something very similar. In general, the violas that clumped well and came back reliably after more than one season tended to be the most popular. Genders listed about forty cultivars in this category.

By the end of the twentieth century, violas had become increasingly popular and diverse. The Princess series was open pollinated, modifying the untidy habit of wild violas such as Johnny-jump-ups. Jagan ("Jaggi") Sharma created the F1 Sorbet series, with its hybrid vigor and range of colors. By crossing specimens from these groups with each other, Sharma came up with vigorous and handsome hybrids.

BELGIUM

Pansies were much admired in Belgium and attracted the attention of important figures. One of these was Louis Van Houtte in Ghent, a mentor of Victor Lemoine and creator of an astounding set of nurseries.

ENGLAND

Lady Mary Elizabeth Bennet (1785–1861), Walton-on-Thames

Lady Mary, the youngest daughter of the Earl of Tankerville, lived on the family estate Mount Felix at Walton-on-Thames in Surrey. She was a child of his second wife. The earl's first wife, Lady Emma Bennet, Countess of Tankerville, had collected exotic plants and had been deeply involved with botany. It was the one intellectual pursuit considered proper for ladies at that time. She employed artists to paint her treasures, and at least one orchid was named for her, *Paphiopetalum tankervillae*. The children who were born at Mount Felix and were interested in flowers found a very congenial environment. Lady Mary's mother was Lord Tankerville's second wife, but she too became fascinated by flowers.

In 1831, Lady Mary married Sir Charles Monck. It was his second marriage, and they had no children. Gardening seems to have filled her time very satisfactorily, and it may have been the thread that pulled them together. As a young woman, she was said to have laid out her own garden and bordered it with pansies collected in the woods. After she married, her gardener saved seed from pansies with the best flowers and began selecting and breeding them.

Very little was known about the work at Mount Felix initially. Horticultural journalists issued frequent bulletins about the progress being made in pansy development during the 1830s, but the work of Lord Gambier's gardener, William Thomson, was better known, and Lady Mary's work was not recognized until somewhat later. The French amateur Ragonot Godefroy published an essay in 1844 giving Lady Mary credit for being the first person ever to create new pansies. A few decades later, Charles Darwin noticed what was going on, referring to Lady Mary in his 1868 *Variation of Animals and Plants under Domestication*.

Wittrock reported that two journalists recorded some thoughts about pansy breeding in 1836: "Chances of success would greatly improve, if one would take the trouble of artificially inseminating the flowers. This would be a suitable task, especially for amateurs, as professional gardeners, whose attention is distracted by other work without time to spare for this, have to rely on the insects carrying the pollen from one flower to another." The authors then explained how easily this can be done by removing the anthers and transferring the pollen with a fine camel hairbrush. These comments make it clear that the religious proscription of intentional pollination had been neutralized by then.

Richard Dean, West Ealing

Richard Dean was a prominent member of the Royal Horticultural Society and secretary of the National Chrysanthemum Society at one point. Dean, a founder of the Royal Gardeners' Orphan Fund, was one of the recipients of the Royal Horticultural Society's Victoria Medal when it was first given in 1897. He also served on the RHS Floral Committee.

Dean introduced 'Blue Bell' in 1872 and told the 1895 Viola Conference it was a chance seedling in his garden at West Ealing. Its presence there was something of a mystery to him, because as far as he knew pansies had not been previously grown in that garden. 'Blue Bell' was a truly blue flower on a neat tufted base.

William Dean (d. 1895), Birmingham

William Dean, who lived in Birmingham, was Richard Dean's brother. He too was devoted to violas and spent much of his career in developing them. He recorded the fact that he began to grow fancy pansies in 1858. William Dean organized a viola conference in Birmingham in 1894 but did not live to attend it. He was in demand as a judge at viola and pansy shows in York, Shrewsbury, and Wolverhampton through his participation in the Midland Pansy Society. The cultivars 'True Blue', 'Bridesmaid', and 'Princess Beatrice' remain his lasting legacy.

James Grieve (1841–1924), Edinburgh

Grieve was born in the old Scottish border town of Peebles and began his apprenticeship at Messrs. Thomas Gentle and Son in that town at the age of twelve. He later moved on to Stobo Castle for four more years of training. In

1859 he started working at the Dickson Nursery in Edinburgh. He was an excellent gardener and rapidly gained more responsibility.

Grieve was the foreman and later the general manager at the Dickson Nursery, working there for almost thirty years. Some still call him "the father of the bedding viola," and he remains a much-admired figure. Graham Hardy, serials librarian at the Royal Botanic Garden Edinburgh, told me that Grieve walked around the botanic garden every New Year's Day for fifty years. Between the 1890s and the 1910s, Grieve was an extramural lecturer at the RBGE, presumably transmitting the benefit of his vast experience to the next generation.

Transforming violas was not his only achievement. Grieve also worked with dianthus and orchard fruit. The "James Grieve apple," particularly suited to the Scottish climate, is very good both for cooking and for dessert, and quite possibly more people know of him as an apple man than as a flower man.

Grieve gave gardeners a new tool for handsome bedding by making numerous crosses between the hybrid show pansies and various species such as *Viola cornuta, V. lutea,* and *V. striata.* The resulting plants were compact and emerged from a mass of fibrous roots, which could be divided easily for propagation. He called these "tufted pansies." They are also known as tufted violas. He took pollen from the most recent of the new show and fancy pansies and applied it systematically to species flowers, many of which he had collected himself in the surrounding countryside.

One exception was *Viola cornuta,* which had come from Spain in 1776. Grieve found that if he reversed the seed parents and pollen source, the resulting flowers resembled the original species rather than being a new departure. Of even greater significance was the fact that some of his cultivars did very well in the southern part of Great Britain.

Other breeders working on similar lines at the same time included Richard Dean. Perhaps the next most striking introduction came from Dr. Charles Stuart in the Scottish borders, whose work resulted in the charming flower he called a "violetta." These too were based on *Viola cornuta,* but the plants turned out to be more compact, with fragrant elongated larger blossoms. Stuart was an amateur who spent ten years improving his cultivars.

When the Dicksons decided to break up the nursery, Grieve and his two sons took over the Redbraes segment of the business and started their own firm. The author of an obituary of Grieve referred to the esteem in which Grieve was held. This was due not only to his vast professional skill and attainments but also to his "genial personal qualities." In 1909, a group of citizens raised a subscription to present him with a gold watch and a bag of gold sovereigns in recognition of his fifty years of hard work and success. The Royal Caledonian Horticultural Society also recognized his achievements by bestowing the Neill Prize on him in 1916.

William Richardson

William Richardson was the Earl of Tankerville's and Lady Mary Elizabeth Bennet's gardener in Surrey. He was a true pioneer. If it were not for the terrible snobbery of the age he would have been feted. We are lucky even to know his name.

Dr. Charles Stuart (1826–1902), Chirnside

Dr. Stuart was a general practitioner in the small town of Chirnside, Berwickshire, in the Scottish Border region. Working with the viola family made a lot of sense because of the cool, moist climate in the Borders. He too used *Viola cornuta* as the seed parent. One of his first crosses was in 1874, using the pollen of 'Blue King', a new bedding type of plant. He wrote about his work in Dobbie's Horticultural Handbook *Pansies, Violas, and Violets* in 1898. "There was a podful of seed which produced twelve plants, which . . . were all blue in colour, but with a good tufted habit. I again took the pollen from a pink garden pansy and fertilised the flowers of my first cross. . . . The seed from this cross gave me more variety in colour of flower, and the same tufted habit of growth."

William Thomson

William Thomson was Lord Gambier's gardener in Buckinghamshire. Just like William Richardson, Thomson laid the foundation of all the work that followed.

FRANCE

Jules Bugnot (1831–1899), St.-Brieuc, Brittany

Bugnot made a big splash at his very first flower show in 1875. No one had ever heard of him before, but he took all the prizes and received extremely favorable comments. This was the third time during my research I have seen such a turn of events. Bugnot was an optician, making his living selling opera glasses and telescopes. He raised pansies as an avocation. In 1889, a representative of the provincial government at Finistère, Charles Albert, the prefect of Côtes-du-Nord, now Côtes-d'Armor's, chef de cabinet, visited him at his home in rue de la Charbonnerie in St.-Brieuc, a high honor indeed. By then, Bugnot was well established as an expert on the pansy, and his pansy seed was in constant demand, but he was not well-known in person. Writing in the bimonthly review *Lyon horticole,* Monsieur Albert heaped praise on Bugnot, holding him up as an object lesson for other people in the field.

The garden was very well located, high over a valley. Albert reported that Bugnot toiled tirelessly in his garden from before dawn to late at night, before and after his normal day's work. He allowed no one to help him, not even his family. Hour after tedious hour he removed slugs and snails and did all the cultivation required to produce prizewinning blossoms. Jules Bugnot was fifty-seven years old at the time of the visit, and it seems disconcerting to us in the modern era that Albert considered that to be a ripe old age. He expressed wonder that Bugnot continued to be so vigorous.

The author did explain how Bugnot had become so interested in growing pansies. Many years before, a visitor to the garden had thought that the situation was ideal for the pansies he was breeding and asked Bugnot to let him have some space for a few of his extras. Until then, Bugnot, like any conscientious family man, grew vegetables in his garden and had a modest number of fruit trees. The arrival of the pansies changed all that. Here was something special he could do, and he began to make crosses. Albert commented that Bugnot had an unerring sense of color and focused particularly on a limpid clear yellow. Like all other great breeders, he knew exactly what he wanted and ruthlessly discarded any plant that was not perfect.

Cassier, Paris

Monsieur Cassier owned a nursery at Suresnes. Later in life he obtained James Odier's collection and advertised them as the Cassier/Odier pansies.

Auguste Miellez (d. 1860), Esquerme, near Lille

Although better known for his roses, Miellez introduced several striking new pansies in the late 1850s. 'Imperatrice Eugenie' was rose-red and white, quite outside the rigid boundaries of the English pansy fanciers' rules, and 'Napoleon III' mainly purple and yellow. His colleagues Charpentier and De May both followed suit. De May's 'Leonidas' was considered to be one of the most elegant pansies ever introduced.

Jacques (James) Antoine Odier (1798–1864), Paris

Jacques Antoine Odier was the son of a distinguished banker, also named Jacques Antoine. The family called the son James, and that is how he became known. Both James Odier and his wife were interested in their garden and depended heavily on their gardener, Jacques Duval, to carry out their ideas about new cultivars. James became regent of the Banque de France and lived in the fashionable seventh arondissement. In addition, they had a country place at Bellevue where they could enjoy the air and do their gardening. They were on some of the land surrounding Madame de Pompadour's chateau, built for her by Louis XV.

Although Odier is mainly renowned for pansies, Duval introduced numerous other prizewinning cultivars, particularly pelargoniums and roses. Many of these were named for members of the family, including the eldest daughter, Louise. Jacques Julien Margottin, a well-known flower breeder, introduced *Rosa* 'Louise Odier' in 1851. Margottin was just as distinguished for his work in chrysanthemums and roses as in pansies. Odier himself concentrated on the central blotch, causing it to be darker and more striking in the pansy cultivars he released.

Alexandre Desiré Trimardeau (1856–1931 [1932?]), Kremlin-Bicêtre

Trimardeau was born in Montoire but settled in Gentilly, the town that became Kremlin-Bicêtre in 1896. His father worked in horticulture, and he followed suit. Alexandre remained at the same address in Kremlin-Bicêtre until his death, probably in 1931. The Trimardeaus had one son, Paul Henri, born in 1886. Tragically, their son died in 1911 at a very young age.

After 1921, the census recorded Alexandre's occupation as landlord rather than horticulturist, as before. By then he was sixty-five and possibly had retired from active work. His unusually large and attractive pansies were introduced in the 1880s to great acclaim. By 1900 Conard and Jones, the American rose specialists, included many French pansies in their catalogue, such as "Improved Giant Trimardeau Pansies." This gives you an idea of how they were sold. The blossoms came in several rich and handsome colors like dark red, deep blue, yellow, and orange. In the succeeding paragraph Conard and Jones offered Cassier's Giant Odier Pansies.

Vilmorin-Andrieux

The venerable firm of Vilmorin-Andrieux also brought out new pansies. These were noted for the size and quality of the blotches.

GERMANY

Wittrock, the Swedish botanist, devotes a section of his viola history to Germany, where there was evidence that pansies were sold very early in the nineteenth century. A Weimar magazine carried a note about a potted dark-violet *Viola tricolor* in 1808. By the end of the 1820s, notes about crossing *V. altaica* with *V. tricolor* to create flowers with large blossoms were appearing in the German literature.

Nurserymen were importing the improved British pansies in the 1830s. Herr Bockmann of Hamburg listed 138 varieties in his 1841 catalogue. Unfortunately, information about many German firms cannot be recovered, because two world wars and the Communist takeover of Eastern Germany led to the loss of many business records.

Ernst Benary, Erfurt

Benary is discussed extensively in *Visions of Loveliness,* and the firm's interest in pansies was well known. In the 1880s, they grew a large number of pansies in Chiswick, at the Royal Horticultural Society's gardens. These included some of the English streaky varieties, and the flowers were widely admired. One blossom, a Bugnot cultivar, was the largest ever seen, about three to four inches across.

Doeppler, Erfurt

Doeppler introduced 'Goldorange', derived from *Viola lutea* Huds. var. *grandiflora* L.

Gotthold and Co., Arnstadt

Gotthold and Co. contributed to the success of pansies in Germany. They worked with the flowers popularized by Salter: yellow and brown with the streaks and red flaming.

A. Knapper, Karlsruhe

Knapper introduced the double-blossomed 'Ottilie von Menzingen' in 1865.

C. Lorenz, Erfurt

Lorenz is remembered for the ultramarine 'Kaiser Wilhelm', with its purple-violet blotches.

Gebrüder Mette (Mette Brothers), Quedlinburg

The city of Quedlinburg in the Hartz Mountain region of Germany had a proud and ancient history as a center of agricultural excellence, an eminence shared with another Thuringian city, Erfurt. The confluence of navigable

Johann Peter Christian Heinrich Mette, founder of the company that became Gebrüder Mette. *Reproduced by permission of Dr. Gerhard Roebbelen*

rivers with surrounding fertile soil had led to settlement early in the Middle Ages and the growth of agriculture. The church was the dominant organization at that time, and both cities started out by being dependent on powerful and wealthy abbeys. Much of that wealth stemmed from the abbeys' crops.

The seed company Gebrüder Mette has a very long history, going back to 1784, when Johann Peter Christian Heinrich Mette (1735–1806), later known simply as Peter Heinrich Mette, founded his business in Quedlinburg. Mette had learned his trade slowly and carefully over several years. Initially he was apprenticed to the Quedlinburg Imperial Abbey's chief gardener, Johan Heinrich Ziemann, between 1750 and 1753. He then traveled to other parts of Germany to gain further experience. By 1784, he was ready to settle down and leased the deanery garden from the local diocese to open his business.

Martin Grasshoff was the first nurseryman in Quedlinburg to offer a major seed collection. In 1792, Peter Heinrich Mette followed suit with his own seed collection. New ideas about agricultural business were coming in fast. At a time when the roads were terrible and much of Europe was at war, distributing plants and seedlings was very difficult, and seeds were far easier to handle. Two years later, Peter Heinrich's son Burghart Hartwig Mette joined the firm, and in 1802 he took over its management.

Burghart was an effective businessman with a lot of initiative. He built several greenhouses and grew exotic crops on a large scale, including citrus of all sorts, jasmine, laurel, oleander, and carob bean. On special occasions, such as royal visits, he gave the abbey unusual fruit for the exalted guests.

Burghart's three grandsons were the reason for the business becoming known as Gebrüder Mette. Helmut Gäde studies the history of agriculture in Quedlinburg and has written two valuable books about its agricultural past. He lists a new scion of the Mette family in each generation until 1945. After that, the business was confiscated by the new Communist government of the Deutsche Demokratische Republik (East Germany).

Johann Georg Heinrich Mette (1867–1907), known as Henry Mette in the English-speaking countries, was active in the last quarter of the nineteenth century and often advertised pansies and violas in the English and American horticultural journals. He offered his own cultivars such as 'Triumph of the Giants', as well as plants developed by the great French breeders Trimardeau, Odier, and Cassier.

The firm prospered mightily, acquiring more land and building large facilities to process seed and do research after 1851. At the peak of their era, more than four thousand varieties of seeds were offered. Improved seed for sugar beets, something in which the great French firm of Vilmorin-Andrieux also excelled, was the foundation for Gebrüder Mette's preeminence. Sugar beets were a crucial European crop when importing sugarcane from the Caribbean Islands and the southern United States was expensive and difficult. Mette was considered to be a leader in its field. Sadly, all that fertile land has been plowed under, and the main site now has a supermarket on it.

Moschkowitz und Siegling, Erfurt

One of Moschkowitz und Siegling's introductions in 1852 was very similar to *Viola tricolor* L. in its coloring, and presumably this species was one of the parents. Several of their other pansies had more unusual coloring. One of them has deeper tones of the same hue in the upper petals compared to the lower ones. They were depicted in a contemporary horticultural magazine. This firm was one of the leaders in its time.

Schwanecke, Oschersleben

In the 1850s, Schwanecke released an almost-black pansy, 'Dr. Faust'. Wittrock noted that it was still available fifty years later. Another of the firm's black cultivars was 'Negurfurst'. One of its parents was the extremely dark blue 'Azurea' of Gebrüder Mette. Its petals were quite satiny and larger than those of 'Dr. Faust'.

H. Wrede, Lüneberg

At this time, Wrede's pansies were very striking. They were illustrated in the floral magazines.

SWEDEN

The pansy grows well in northern latitudes and was popular in both Sweden and Norway during that epoch. A few flowered at latitude 70 degrees north. Very unusual seeds were produced at the Heimli farm in Tanen, able to survive the winter under snow. V. B. Wittrock's thorough and exhaustive studies on the pansy and viola have not been surpassed.

UNITED STATES

W. Atlee Burpee took note of the great development going on and sent one of his agents to meet with some English experts in 1886. A year later, a large order of pansy seed arrived in Philadelphia. The plant, renamed 'Burpee's Defiance Pansies', was promoted in the 1888 catalogue, although Burpee did not reveal the name of his source. From the surviving pictures, Wittrock suggests that they were of French or German origin. A few years later, Peter Henderson in New York also listed pansies in his catalogues. These were clearly British sweet-scented violas from the illustrations. Henderson did also breed some of his own, among them 'Henderson's Mammoth Butterfly Pansies'.

8

Water Lily

LUSHLY ROMANTIC, often blooming at night with an exquisite fragrance, water lilies are a great source of delight. They were esteemed by many previous civilizations for the same reasons. Perhaps the one that resonates the most with us is Egypt. The images of lotuses are inextricably entwined with those of pyramids and paintings of individuals in profile. Curiously enough, however, Sir Francis Bacon, author of *On Gardens* in 1625, did not care for these flowers. The *Gardener's Chronicle* in the mid-nineteenth century also dismissed the plant, as did the redoubtable William Robinson.

Members of the genus *Nymphaea* may be found in temperate and tropical conditions throughout the world. Perhaps the most astonishing species is *Victoria amazonica*, first seen by European explorers in Bolivia. The leaf can reach up to nine feet in diameter, supported by an extraordinary, almost architectural system of buttressing on the underside of the pad. It is one more example of the superb tensile strength achieved by a plant in what seems to be a hopelessly fragile structure. Most people have seen the pictures of small

George Pring.
Reproduced by permission of Missouri Botanical Garden

children standing on the giant pad. One former child model has left a record of how it actually felt.

The daughter of George H. Pring, one of the twentieth century's most prominent experts in water lilies, Isabella Seibert, recounts how her father posed her on one such leaf with an exhortation to smile, as well as many reassurances that she would be fine. Said daughter felt herself wobbling around and was not quite so certain.

BOTANY

There are three genera: *Nymphaea, Nelumbo,* and *Victoria.*

Nymphaea

At one time, all water lilies were considered to be in the same botanical family, Nymphaeaceae, but recent molecular research has shown that the water lily and the lotus have different gene sequences in the chloroplasts, and lotus is now in the family Nelumbonaceae. Although there are external similarities, the plants may still be distinguished by their flowers, leaves, and rhizomes.

Nymphaeaceae is subdivided in several ways. According to Perry D. Slocum, it has seven genera: *Nymphaea, Nuphar, Nelumbo, Victoria, Euryale, Ondinea,* and *Barclaya* (syn. *Hydrostemma*). Most hybridizing takes place in the *Nuphar* and *Nymphaea* genera. Each genus of *Nymphaea* has various subgenera. These in turn have many species, of which three have principally been used in hybridizing: *Castalia* for hardy water lilies, *Lotos* for night-blooming tropical water lilies, and *Brachyceras* for day-blooming tropical plants. Chinese propagators have made countless hybrids of *Nelumbo* for centuries. These hybrids are available in other countries, including the United States.

The more useful classification of *Nymphaea* for horticultural purposes is that between tropical and hardy types. The names indicate the source of the differences, primarily via geography. Hardy water lilies can be found in the Northern Hemisphere, including the United Kingdom, France, Sweden, North America, and northern Asia. Their flowers bloom during the day. *Nymphaea alba* and its varieties are the most commonly hybridized in this group, with *N. odorata* close behind. The diminutive *N. tetragona* has also been used extensively. It produces seed very prolifically.

Within the tropical group, some flowers bloom and emit their fragrance in the daytime and others bloom and become fragrant only at night. Five principal species are used most frequently for residential water gardens: *Nymphaea capensis, N. colorata, N. gigantea, N. lotus,* and *N. rubra.*

Hybridizing water lilies for the temperate zones started in the nineteenth century. Joseph Paxton claimed to be the first to do this in 1851, but there were doubts from the start that his alleged cultivar, *N. devoniensis*, supposedly a cross between *N. rubra* and *N. lotus,* was more than a form of *N. rubra*. The name was in honor of his august employer, the Duke of Devonshire.

What is not in doubt is the work of Eduard Ortgies, a gifted German horticulturist employed first by the duke to get *Victoria amazonica* to flower in Europe under glass and later by Louis Van Houtte. Ortgies crossed *Nymphaea ortgesiana* with *N. rubra* to derive *N.* 'Ortgesiana-rubra', a tropical night bloomer. Ortgies is said to have created other hybrids, but there are no reliable reports of them.

Once he became the head gardener at the Zurich Botanical Garden, he bred no more water lilies. He made no secret of his methods, however, and probably laid down the path that others were to follow. This was reported by J. E. Planchon, a French botanist who had worked at Van Houtte's school of horticulture in Ghent.

The Inspector of the Berlin Botanical Garden (now at the Berlin-Dahlem Botanical Garden), Claude Bouché, is the next person who is reliably known to have hybridized new cultivars of water lilies. It seems likely that he visited the Van Houtte establishment in Ghent. Bouché had access to an unlimited supply of warm water, effluent from an iron foundry, and was able to grow his new plants outdoors. In 1852 and 1853 he introduced several new tropical night-blooming hybrids. He fertilized *N. rubra* with *N. lotus* pollen and obtained seven cultivars. Another nine were the result of pollinating the new hybrids with *N. lotus* again.

Thirty years elapsed before another authentic new cultivar appeared, this time from the Royal Botanic Gardens at Kew. A gardener named Watson pollinated *N. lotus* with *N. devoniensis* in 1885. The resulting *N. kewensis* was reported in *Gardener's Chronicle* in 1887 and in the *Botanical Magazine* in 1888.

Nelumbo

There are two species in this genus: *Nelumbo lutea* and *N. nucifera*. *N. lutea* may be found in the eastern and central United States. *N. nucifera* is endemic

Nymphaea coerulea 'Blue Lotus of the Nile'.
Shutterstock.com

to Asia, the Philippines, northern Australia, Egypt (though there is a possibility that it was taken to Egypt from India about 500 BCE and is not truly endemic to Africa), and the Volga River basin in Russia. Unfortunately two of the flowers popularly called lotus do not belong to this family. 'Blue Lotus of the Nile' is not a lotus but probably *Nymphaea coerulea,* and the 'Blue Lotus of India' is *Nymphaea nouchali* (syn. *N. stellata*).

Victoria

This genus has two known species: *Victoria amazonica* (formerly called *V. regia*) and *V. cruziana*. The first specimen of *V. amazonica* was collected by the Czech botanist Thaddäus Haenke in Bolivia. He sent it to Europe in 1801. Subsequently the French botanist Aimé Bonpland found it in Corrientes, Argentina, in 1819, and Eduard Friedrich Poeppig found one in the Amazon in 1832. He named it *Euryale amazonica.*

Alcide d'Orbigny collected and recorded the second species, *V. cruziana*, also in Corrientes in 1840. Its leaves are a slightly lighter green. Most usefully, this species tolerates somewhat cooler temperatures, allowing Europeans to grow it outdoors.

The first set of *V. amazonica* seeds sent to Mr Bridges, a gardener at Kew, from Bolivia germinated, but the resulting plants did not survive. A few years later, Dr. Rodie and Dr. Luckie, physician explorers, sent seeds sealed in a bottle of water to Joseph Paxton at Chatsworth. This worked. Those seeds gave rise to healthy plants, and on November 8, 1849, the first flowers appeared. Paxton shared his seeds with colleagues on the Continent and in the United States. The flower of Caleb Cope's *Victoria regia* opened in Philadelphia in August 1852.

BREEDERS

BRITAIN

Peter Bisset (1869–1951), Auchtermuchty, Fife

Bisset was born in Auchtermuchty in Scotland, and he learned a good deal of what he later applied to his work from his forester father. Bisset was an authority on the culture of the water lily and wrote a book initially published in 1909 and reissued twice in the next twenty years. He crossed *Nymphaea dentata* with *N. sturtevanti* to create *N. bisseti*. *N. dentata* was a parent of the rosy red *N.* 'O'Marana'. Another seedling was *N.* 'George Huster', which had a crimson color.

Ray Davies, Stapeley Water Gardens

In addition to the Stapeley Water Gardens, founded in 1965 near Nantwich in Cheshire in England, Davies also owned the Latour-Marliac establishment in France. Ray Davies worked with his wife, Barbara. The nursery was closed in 2011 after a decline in business. The land became more valuable for development.

James Gurney (1831–1920)

Like so many other men in this chronicle, James Gurney was the son of a gardener. He had been born in Buckinghamshire and went to work at the Royal Botanical Gardens, Kew, at an early age. One of his duties was to superintend the growing and flowering of *Victoria amazonica.*

Amos Perry (1871–1953)

Perry was noted for his lilies and hemerocallis but also wanted to breed new water lilies. He owned a successful nursery, the Hardy Plant Farm in Enfield, and was always looking for new plants to increase his stock. He had visited Joseph Marliac (see below) in France but came away without any useful advice.

Working on his own in the 1930s, Perry bought *Nymphaea tetragona* (then called *N. pygmaea*) seed from T. Smith's Daisy Hill Nursery in Newry, Northern Ireland. He bred many fine cultivars and won awards at the Royal Horticultural Society shows, but he found it hard going, because the flowers often failed to set seed. He was very frustrated in one season when only one seedpod ripened from more than 150 crosses.

Amos Perry was not the only distinguished plant person in Enfield. E. A. Bowles, an expert on small bulbs and avid creator of new cultivars, also lived and worked in that small town, now absorbed into Greater London. It is intriguing to wonder whether they knew each other.

FRANCE

Joseph Bory Latour-Marliac (1830–1911)

Latour-Marliac was the scion of a cultured family near Bordeaux in southern France, with several relatives being actively involved in the sciences. His father was a botanist and grew many unusual types of trees on the family estate. By about 1860, bamboo, Japanese persimmon *(Diospyros kaki),* and eucalyptus were among them. The young Marliac studied law in Paris, but the 1848

Joseph Latour-Marliac.

Communist uprisings in Paris interrupted his education, and he returned home to manage the estate. After marrying, he settled in Le Temple-sur-Lot, just a few miles from the family property.

His passion for water plants emerged at the end of the 1850s. An American expert, Kit Knott, notes that a considerable amount of myth has grown up around these activities. As in so many other situations, possibly no one will ever really know what happened. Somehow Marliac came across brilliantly colored tropical water lilies and thought it would be wonderful if he could create hardy water lilies in the same hues. It is that same old "vision" thing popping up once more. The native hardy water lilies in his region were only white or very pale pink.

The estate had three large natural springs, which were siphoned into forty ponds. These in turn were subdivided into six hundred channels. This system permitted him to grow dozens of new plants in order to find the few that would be worthwhile. Joseph also added a very large greenhouse to the estate to further his experiments.

Nymphaea alba var. *rubra.*
Shutterstock.com

Marliac was aware of the white *Nymphaea alba* widespread across Europe. In 1878 a pink sport was introduced, *N. alba* var. *rubra.* With age, the flower turned an even deeper pink. It had been found on a Swedish lake by the Swiss plantsman Otto Froebel.

A contemporary article commented on a canary-yellow water lily found in Florida, later named *N. mexicana* (syn. *N. flava*). Marliac is thought to have used *Nymphaea odorata* L. and some of its colored variants from the United States as one of the parents for his crosses. William Hamilton took this plant, the yellow variant of *N. odorata,* to England in 1786 and it was the first of its kind there.

The variety *N. odorata* var. *rosea* came from a desiccated lakebed or pond in West Virginia, the existence of which had been previously unknown to the people living there.

Joseph Marliac was a secretive man and enjoyed mystifying people. No one really knows precisely how he did his crossings (see Strawn discussion below). When Amos Perry visited him from England in the early 1900s,

Perry was expert with daylilies and other genera but also tried his hand at *Nymphaea*. Marliac told him he would sell Perry his secrets for a thousand pounds. If Perry did not pay, he would have to wait forty years to learn about them. In the event, nothing of the sort took place. To state the obvious, Marliac's secrets died with him.

He left a legacy of about seventy-five cultivars, some of which are still in commerce. 'Conqueror' (red and white), 'Ellisiana' (purple-red), 'Fulva' (coppery red), and 'Gladstone' (white) won the Awards of Merit. After Marliac died, his firm continued to introduce new cultivars of *Nymphaea* until the Second World War.

Marliac must have liked England and the English, in spite of teasing Amos Perry. He regularly submitted his new water lilies to the Royal Horticultural Society's shows and often won Awards of Merit for them, a high honor. Among other things, this helped to increase sales.

One of the most charming resonances between horticulture and art lay in Claude Monet ordering water lilies from the firm of Latour-Marliac in 1894 and again in 1904. The invoices have been preserved, and one can see what Monet wanted to get his brand new water garden going. With the water lilies growing vigorously in the ponds, Monet proceeded to paint a long series of pictures emphasizing the flowers' soft romantic beauty. Curiously, not all the water lilies Monet bought were bred by Marliac. Several came from Dreer in Philadelphia.

SWITZERLAND

Otto Froebel (1844–1906), Zurich

Froebel worked in Zurich. His father, Theodor, who had been a gardener, had founded a landscape architecture business. Father and son continued to maintain the nursery and develop new varieties of plants while they designed and installed new gardens. This base informed their work and led to it being more plant centered rather than driven by principles of abstract design.

Otto spent forty years working with *Nymphaea alba* var. *rubra*. He had very rigorous standards of selection. The handsome water lily known as

'Froebeli' is not a hybrid but a variant of *N. alba* var. *rubra*. It has very dark wine-red petals and vermilion stamens. 'Froebeli' was still in commerce at least until recently.

THAILAND

Slearmlarp Wasuwat (d. 2014)

Wasuwat is known as the "father of the water lily" in Thailand. (His friends and students also affectionately referred to him as "Sam.") He introduced many new hybrids and gave talks and demonstrations across the country for years. Numerous young horticulturists learned about water lilies from him, spreading the word and creating wide interest.

He introduced more than three hundred water lilies in thirty years, some collected in the field and others he bred himself. Among his favorites were *N.* 'Nangkwang Chapoo 1', *N.* 'Primlarp', and *N.* 'Gulyanee'. The vibrant pink *N.* 'Larp Prasert' was one of his earliest efforts, and it remained in commerce until quite recently. To press home his points, he wrote six books on the subject.

UNITED STATES

A great deal of work was going on in the United States at the end of the nineteenth century, though most of the new introductions were the result of chance sports or the pollinating activities of insects rather than human intervention. Enough new flowers were introduced during that epoch that Edmund T. Sturtevant, in Bordentown, New Jersey, issued a catalogue of water plants in 1881. It was said to be the first of its kind. Sturtevant was fortunate in owning a sheltered millpond and so was able to get even quite tender plants to bloom in quantity.

He received seeds of *Victoria amazonica* from Edward S. Rand Jr., in Para, Brazil. The flowers started out white but turned crimson over time. Sturtevant found that slashing the seeds improved the rate of germination. James Clark in nearby Riverton had noticed the same thing.

Henry S. Conard (1874–1971)

Conard taught botany at Grinnell College in Iowa for almost fifty years. He was expert in both water lilies and the mosses, and was also an early proponent of ecology and the importance of the environment. In 1905 he published one of his major works, *The Water-Lilies: A Monograph of the Genus Nymphaea*. It remains a standard reference for water lily science even today.

Conard had been born in Philadelphia and educated in leading establishments on the East Coast, but in 1906 he left Johns Hopkins, where he was teaching botany, to take the chair in botany at Grinnell. Even after officially retiring from Grinnell in 1944, Henry Conard continued to write and edit almost until his death in Florida at the age of ninety-seven.

Emmett, Anne, and Verena Liecht, Jim's Watergardens

The Liechts were very well known in their day and made significant contributions to the field.

Joseph Lingg, Ardsley, New York

Lingg's Aquatic Gardens was on one of the main thoroughfares in Ardsley. One of his cultivars, 'Crystal Lingg', was named for his mother.

Rolf Nelson, Key, Texas

Rolf Nelson worked very closely with his wife, Anita. Both were graduates in horticulture, and Rolf had worked for the Thomases at Lilypons in Maryland for many years before deciding to open his own business. They found suitable property in the Houston area and started their nursery in 1997. One of his cultivars is 'Texan Shell', a tropical night-blooming flower with creamy white and purple coloring.

George H. Pring (1885–1974)

Pring was the youngest person ever to graduate from the horticultural school at the Royal Botanic Gardens at Kew. He was recruited to work on the orchid collection at the Missouri Botanical Garden in 1906, but was quickly seduced by the beauty of water lilies and switched to them instead. Apart from their intrinsic beauty, Pring found that they yielded useful new hybrids much more quickly than orchids. Instead of having to wait seven years, he could see success in one year.

This does not mean that he gave up orchids. During his tenure at the Missouri Botanical Garden, from 1910 to 1963, he continued to excel in orchid breeding and organized many valuable shows to encourage others to excel too. For fifty years, Pring was in charge of horticulture at the garden under a succession of directors.

He particularly enjoyed working with tropical water lilies and made numerous trips to collect them in the wild. Pring introduced about thirty tropical day-blooming cultivars in many hues. He named them for his family, for important persons, and many in response to their color and beauty. The most poignant is the one he named for his son, killed in action in the First World War, the deep yellow 'Aviator Pring'. Pring left a legacy of mainly tropical water lilies, with a few hardy types. The very first patent for a hybrid water lily was granted to Pring in February 1933 for *N.* 'St Louis' (following passage of the Plant Patent Act of 1930). It was plant patent no. 55. The law had only gone into effect in 1930, and it is unclear whether the patent belonged to him personally or to his employer, the Missouri Botanical Garden.

Martin R. Randig (1897–1967), San Bernardino, California

Randig lived and worked in San Bernardino, California, working very largely alone and quite anonymously as far as his neighbors were concerned. Almost no one knew that he had ponds filled with exquisite new water lily cultivars behind the nondescript walls of his modest home. The blossoms of Randig's water lilies had a finer fragrance and more unusual colors than any previously seen.

Randig was a very cautious man, but he responded quickly when Ted Uber of Van Ness Nurseries asked him to work with the firm. With this opportunity, Randig finally had a reliable place to which he could consign his cultivars and gain some recognition. Randig received the second patent for a water lily in 1938. The cultivar was named for his wife.

The story of Ted Uber (1908–2007) is unusual. He was a master aluminum craftsman with a thriving business in metal objects for the house when the Korean War reduced the supply of aluminum for nonmilitary purposes to a trickle. In order to survive, Ted became a teacher in industrial arts at a local high school for a time. Things looked up when his cousin Edith Van Ness asked him to take over her water lily business in Upland, California. He accepted. The supply of aluminum subsequently returned to prewar levels, and he picked up his former work again, alternating between the two lines of business.

Randig consigned amazing new flowers to Van Ness, such as 'Afterglow' and 'Green Smoke'. Gardening reporters quickly found out about them and featured them in magazine articles. This boosted sales very effectively. About fifteen cultivars can be attributed to Randig. Uber's son William also bred water lilies of his own.

Perry Slocum (1913–2004), Winter Haven, Florida

Slocum was born near Cortland in upstate New York. He attended Cornell University where he was pre-med, but gave that up once he became fascinated by water lilies. He and his wife, Trudy, set up Slocum Water Gardens first in Marathon, New York, and then in Binghampton, but subsequently they moved to Winter Haven, Florida. Chilly upstate New York is less conducive to growing water lilies than the much warmer Florida.

Slocum was a noted photographer but he also introduced several very fine new cultivars of *Nymphaea*. He obtained the first patent ever issued for a hardy water lily, *N*. 'Pearl of the Pool', in 1946. In 1986, he received three more patents, also firsts, for *Nelumbo* 'Angel Wings', 'Maggie Belle Slocum', and 'Charles Thomas'. Slocum wrote useful books about water lilies and lotuses. Slocum and, later, Charles B. Thomas both received patents right after Pring and Randig.

In between these special events, Slocum introduced about fifty handsome cultivars. At one point Slocum "retired," but he very quickly set up a new business in Franklin, North Carolina. After his death, his stepson Ben Gibson continued the business as "Perry's Water Gardens."

Robert Kirk Strawn (1922–2008), College Station, Texas

Strawn did a great service to the field of water lily gardening by sorting out the myriad overlapping of names and mistaken identity of plants sold commercially. Strawn was born in Florida but spent the larger part of his life at Texas A and M University in College Station. He moved there to teach biology and ichthyology but became intrigued by his wife Charlene's hobby of growing water lilies.

He decided he would collect specimens of every water lily available and watch how they grew. In this way, he could identify different cultivars that were not identical but had the same name. In the reverse situation, some cultivars that had been given different names turned out to be identical. (This problem is not unique to water lilies but is often found in many other genera. I am aware of it in the olive tree, and have seen it discussed in the peony fraternity.)

When Strawn began examining the Marliac cultivars in his collection, he saw that none of them was fertile. This was a very significant discovery, and explained things about the Marliac output that had been previously observed. The question was whether Marliac had done this intentionally, and if so, why. The consensus based on documents Marliac left is that he had indeed intended to do this, and probably for two reasons. One was to maintain the purity of his lines to prevent people from messing about with them and perhaps issuing compromised plants under his name. The other was economic. It meant no one else could make money from his cultivars without his permission.

Strawn then set about breeding his own water lilies. By a quirk of chance, the first one he introduced had been pollinated by insects rather than by his agency. Insect pollination is seldom a reliable method of hybridizing in water lilies. He named the first result for his wife, *Nymphaea* 'Charlene Strawn'. Before he was done, he had introduced more than fifty cultivars. They were available at the Strawn Water Gardens. Like so many other men of extraordinary energy and talents, he was not content to work solely on one type of plant. He also put a lot of time and effort into the iris and created even more cultivars in that genus. He reminds one of the amazing Frank Reinelt (see my previous volume, *Visions of Loveliness*), who changed the world of begonias and also of delphiniums.

George Leicester ("Lester") Thomas (1880–1969), Frederick, Maryland

Thomas began his Three Springs Fisheries water gardening and goldfish business on his own land near Adamstown, south of Frederick, Maryland, in 1917. Thomas had earned a master of arts degree at Franklin and Marshall College in Lancaster, Pennsylvania, and followed various pursuits before finally developing the water garden business. At one point he was the principal of a school in Hallstead, Pennsylvania.

Lester's new enterprise flourished, and the volume of mail orders became so heavy that in 1935 the U.S. Post Office decided to open a new post office in southern Frederick to handle the increasing load. They asked Thomas to suggest a name for the new branch. The name he chose was perhaps a little unexpected but very apt. The coloratura soprano Lily Pons was the great star of the day. Because George enjoyed opera and was one of her fans, he suggested that they name the place Lily Pons, lightly playing on the lily ponds at the nursery.

The unimaginative if not actually Philistine officials at the postal service had other thoughts. They preferred to use a single word as the name, for reasons which are unknown. Many towns and villages in the United States have two-word names, such as White Plains, New York, or Blowing Rock, North Carolina, but the postal authorities insisted on it becoming "Lilypons." It has remained Lilypons ever since, and Charles B. Thomas renamed the nursery to match it in 1977.

George Leicester Thomas Jr. ("G.L.") (1907–1977) was also graduated from Franklin and Marshall College, and joined his father in the business. "G.L." perfected the method of sending fish safely through the mail in polyethylene bags inflated with oxygen. His son Charles took over the firm in 1975, and now Charles's eldest daughter, Margaret Mary Thomas Koogle, is in charge. Charles founded the International Water Lily and Water Garden Society in 1984.

Lester named one of his first cultivars, *Nymphaea* 'Mrs. C. W. Thomas', for his mother and another one, *N*. 'Dorothy Lamour', after the popular movie star. *N*. 'Margaret Mary' was named for his eldest granddaughter.

William Tricker (1853–1916)

The firm of William Tricker is still very active today in Independence, Ohio, though it is no longer in family hands. The first William, whose complete name was Charles William Brett Tricker, was born into a gardener's family in England and became an apprentice at the Royal Botanic Gardens in Kew. When he immigrated to the United States, he initially worked for Henry Dreer in Philadelphia. He eventually established two aquatic nurseries of his own in New Jersey, the first one in 1892. Tricker's youngest son, Charles, took over the business after William died.

William Tricker became prominent when he differentiated *Victoria cruziana* from *V. amazonica* in the early 1890s. In 1897, he published a book about the care and culture of water lilies: *The Water Garden*. Perry Slocum commented that he had learned how to raise water lilies from reading this comprehensive and authoritative book.

Tricker is known for several excellent cultivars. One was *Nymphaea* 'Blue Beauty', a name chosen by the American Committee on Nomenclature because it was considered to be identical with Henry Conard's blue 'Pennsylvania', introduced at about the same time. Tricker had named his cultivar 'Pulcherrima'.

AUTHOR'S NOTE

Collecting all the information presented in these pages has been an unmitigated joy, and it has left me with a great sense of clarifying history and giving wonderful people their due. We who enjoy the result of so many painstaking efforts (often unthinkingly) can now look at familiar plants in a new light. Instead of being merely humdrum, those dull old bedding plants have a distinguished ancestry that deserves to be better known. The knowledge increases the pleasure.

ACKNOWLEDGMENTS

I am indebted to Brian Young of Nottingham, United Kingdom, who generously reviewed the manuscript in its formative stages with an eagle eye.

Chapter 1: Poinsettia

Professor Jules Janick, Roberto G. Lopez, and Christopher J. Currey of the Department of Horticulture at Purdue University all contributed to this chapter in an earlier presentation.
Axel Borg, state enological librarian, University of California at Davis
Diodora Bucur, Ottawa, Canada
Lyndon Drewlow, retired breeder of poinsettia, California
Susan McGregor Epstein, Charleston Horticultural Society, Charleston, South Carolina
Sarah Fick, Charleston Preservation Society, Charleston, South Carolina
Joel T. Fry, curator, Bartram's Garden, Philadelphia
Judith Hines, historian, Charleston Horticultural Society, Charleston, South Carolina
Norma Kobzina (deceased), science librarian, University of California at Berkeley
H. Walter Lack, director, Berlin-Dahlem Botanical Garden
Elizabeth MacLeod, Wilson Library, University of North Carolina, Chapel Hill
Patti McGee, Charleston Horticultural Society, Charleston, South Carolina
Christina Shedlock, Charleston County Public Library, Charleston, South Carolina
Freek Vrugtman, International Lilac Registrar, Hamilton, Ontario

Chapter 2: Chrysanthemum

David Abbott, horticulturist, Chichester, West Sussex, United Kingdom
Monique Barrier, historian, Valence, France
Ann-Sophie Berthon, Paris, France, archivist, Société Nationale d'Horticulture Français
Thomas Brown (deceased), landscape historian, Petaluma, California
Eileen Colwell, local historian, Chichester, West Sussex, United Kingdom
Julie Delavie, Grenoble, France
Julie Dujardin, Paris, France
Maureen Horn, librarian, Massachusetts Horticultural Society, Wellesley

Michel Javaux, Lyon, France
Odile Masquelier, Lyon, France
Josette Sakakini, Marseilles, France
Henri Tachoire, Marseilles, France

Chapter 3: Penstemon

Thomas Brown (deceased), horticultural historian, Petaluma, California
Jean-François Gonot, Lemoine Museum, Nancy, France
Dale Lindgren, University of Nebraska–Lincoln Extension
Zilla Oddy, archivist, Hawick, Scotland
George Thorburn, horticultural specialist, Broadway, Worcestershire, United Kingdom

Chapter 4: Gladiolus

Thomas Brown (deceased), landscape historian, Petaluma, California
Nigel Coe, librarian, British Gladiolus Society
Denis Diagre, archivist, National Botanic Garden of Belgium
Zynara Gaerlan, librarian, Simcoe, Ontario
Leelyn Johnson, librarian, Library of Michigan
Midge Kjome, archivist, Decorah Genealogical Society, Decorah, Iowa
Sharon Klein, town historian, Berlin, New York
John Mahy, nurseryman, Guernsey, United Kingdom
Barb Petroski, Portage County Historical Society, Ravenna, Ohio
Klaus Pfitzer, nurseryman, Fellbach, Germany

Chapter 5: Dianthus

Jennifer Harmer, hardy plant specialist, Winchester, United Kingdom
Michel Javaux, president, Lyon Horticole, Lyon, France
Charles Johnson, librarian, Ventura County, California
Alan Leslie, International Dianthus Registrar, Royal Horticultural Society, London
Connie Piazza, reference librarian, Joliet Public Library, Joliet, Illinois
Douglas Thompson, reference librarian, Redondo Beach, California

Chapter 6: Clematis

Nic Beale, East Grinstead Society, East Grinstead, United Kingdom
Linda Beutler, curator, Rogerson Collections, Oregon
Rosine Chambrillon, Société Horticole Vigneronne et Forestière de l'Aube, Troyes, France

Duncan Donald, International Clematis Registrar, Royal Horticultural Society
Maurice Horn, Joy Creek Nursery, Oregon
Maria Kaczmarczyk, research scholar, Poland
Szczepan Marczyński, clematis breeder, Poland
Elizabeth Mundlak, author's cousin
Ewa Szulc, blog master, Poland
Ken Woolfenden, editor and webperson, International Clematis Society

Chapter 7: Pansy/Viola

Graham Hardy, serials librarian, Royal Botanic Garden, Edinburgh
Alexandra Holmgren, translator, Sweden
Lars-Gunnar Reinhammer, curator, Bergius Botanic Garden, Stockholm
Emmanuelle Royon, Société Nationale d'Horticulture Française, Paris
John Snocken, National Viola and Pansy Society, United Kingdom
Graham Wells, National Viola and Pansy Society, United Kingdom

Chapter 8: Water Lily

Charles Thomas, Lilypons Water Gardens, now residing in Waynesboro, Pennsylvania

NOTES

Chapter 1: Poinsettia

6 "The phytoplasma was later shown...": Lee 1998.
8 "a new *Euphorbia*...": Fry 1994.
9 "Dr. H. Walter Lack...": Lack 2011, personal communication.
9 "Mr. Poinsett was rewarded...being named after him": Stille 1888, 407–8.
11 "Mrs. Herbemont [was] *very vexed*...": Fry 1994 (emphasis in original).
13 "Major advances...": Post 1942, 5, 10.

Chapter 2: Chrysanthemum

25 "Sabine's list of chrysanthemums...": Genders 1971.
31 "displayed single-flowered chrysanthemums...": *Garden*, 17 November 1900.
33 "bronze sport...": *Gardening Illustrated*, 9 November 1895, 556.
35 "a handsome Japanese incurved flower...": *Garden*, 27 October 1894, 384.
38–39 "We have no chrysanthemum society...": Morton 1891, 20.
51 "few plants afford more gratification...": Hovey 1846, 216.
51 "the coming flower": S. L. Emsweller et al. 1937, quoted in Anderson 2007, 391.
54 "Condon...was mentioned...": Miller 1898, 685.
54 "being first for best twenty-four...": Sargent 1893.
54–55 "There is no commercial house... 'Mrs. Alpheus Hardy'": Morton 1891, 37.
57 "This year [1888] Messrs. Hill & Co....": Ibid., 35.
58 "such a name as that of John Lane...": Ibid., 40.
58 "The enterprise...of Messrs. Pitcher & Manda,...": Ibid., 37.
59 "long hold an important place...": Ibid., 36.
59 "of robust habit...": Ibid.
60 "There are also many other excellent varieties...": Ibid., 37.
60 "The progress of chrysanthemum growing...": Ibid., 38.
60 "The name of John Thorpe...": Ibid.
62 "In 1883, Mr. H. Waterer,...": Ibid., 35.
62 "disseminating the productions...": Ibid., 34.
62–63 "the fifty dollar silver cup...Mr. Harris": Ibid., 35.

Chapter 3: Penstemon

69 "gorgeous garden fatties": Farrer 1919, 50.

Chapter 4: Gladiolus

96 "E. S. Rand, Jr., as chairman…": Beal 1916, 156.
96 "was a pure white…": Ibid., 157.
96 "was of the reverse-flowered form…": Ibid.
97 "a spike of gladiolus Calypso…": Massachusetts Horticultural Society 1880, 328.
107 "The neighbors and competitors…": Clarke Historical Library, Central Michigan University.

Chapter 5: Dianthus

120 "Levi Lamborn…": Lamborn 1901, 23.
120 "According to Lamborn…": Ibid., 22.
121 "reliable and painstaking": Ibid., 70.
133 "They should quietly sleep…": Ibid., 24.
133 "as gleeful… to his club": Skidelsky 1916, 37–38.
140 "one of the most successful… extensive business": Futhey and Cope, 2:729.

Chapter 6: Clematis

152 "There was no room…": Ken Woolfenden.

Chapter 7: Pansy/Viola

172 "branched honey stripes…": Wittrock 1895.
177–78 "Chances of success…": quoted in Wittrock 1895, 8.
178 "Dean introduced 'Blue Bell'": Jordan et al. 1898, 25.
180 "There was a podful of seed…": Ibid., 20.

REFERENCES

Chapter 1: Poinsettia

Anderson, Christine, and Terry Tischer. 1997. *Poinsettias: Myth and Legend.* Tiburon, CA: Waters Edge Press.

Britton, Nathaniel Lord, and Addison Brown. 1913. *An Illustrated Flora of the Northern United States, Canada and the British Possessions.* Vol. 2. New York: Charles Scribner's Sons.

Bucur, Diodora. 2010. "Holiday Poinsettias Are Native to Mexico." *Orléans Star* (Ottawa), December 22. www.orleansstar.ca.

Calderon de la Barca, Frances. 1843. *Life in Mexico.* Reprint, Garden City, NY: Doubleday, 1966.

Ecke, Paul, III, James E. Faust, Jack Williams, and Andy Higgins. 2004. *Ecke Poinsettia Manual.* Batavia, IL: Ball Publications.

Fry, Joel T. 1994. "The Introduction of the Poinsettia at Bartram's Garden." *Bartram Broadside* (Winter): 3–7.

Fuentes Mares, José. 1984. *Poinsett: Historia de una gran intriga.* 7th ed. Mexico City: Oceano.

Lee, Ing-Ming. 1998. "The Secret of Free-Branching Poinsettias." *Agricultural Research* 46 (12). https://agresearchmag.ars.usda.gov/1998/dec/poin/.

Longacre, James B., and James Herring. 1837. *The National Portrait Gallery of Distinguished Americans.* Vol. 1. Philadelphia: James B. Longacre.

Miranda, Justino. 2004. "Mexican Authorities Looking to Produce Variant of Flower Native to Morelos State, Circumventing Poinsett's Legacy." *El Universal,* December 20.

Park, Ki-Ryong, and Robert K. Jansen. 2007. "A Phylogeny of Euphorbieae Subtribe Euphorbiinae." *Journal of Plant Biology* 50 (6): 644–49.

Poinsett, Joel Roberts. 1824. *Notes on Mexico.* Philadelphia: H. C. Carey and I. Lea.

Post, Kenneth. 1942. "Effects of Daylength and Temperature on Growth and Flowering of Some Florist Crops." *Cornell University Agricultural Experiment Station Bulletin,* no. 787 (June): 1–70.

Prescott, William Hickling. 1843. *The History of the Conquest of Mexico.* New York: Harper and Brothers.

Rippy, J. Fred. 1935. *Joel R. Poinsett, Versatile American.* Chapel Hill, NC: Duke University Press.

Stille, Charles. 1888. "A Biographical Sketch of the Hon. Joel R. Poinsett, of South Carolina." *1887 Charleston Year Book,* 380–424. Charleston, SC: Lucas, Richardson.

Chapter 2: Chrysanthemum

Anderson, Neil O., ed. 2007. *Flower Breeding and Genetics: Issues, Challenges and Opportunities for the 21st Century*. Dordrecht, Netherlands: Springer.

Bailey, L. H. 1914. *The Standard Cyclopedia of Horticulture*. New York: Macmillan.

"Boucharlat, Joseph, Obituary, 1895." 1916–17. *Revue horticole* 7:294.

Burbidge, F. W. 1884. *The Chrysanthemum: Its History, Culture, Classification, and Nomenclature*. London: "The Garden" Office.

Carrière, E.-A. 1891. "Le premier semeur en France, de chrysanthèmes de la Chine." *Revue horticole* 63:201–3.

Duthie, Ruth E. 1984. *English Florists' Flowers and Societies*. Shire Garden History. Aylesbury, Buckinghamshire: Shire Publications.

———. 1988a. *Florists' Flowers and Societies*. Shire Garden History. Aylesbury, Buckinghamshire: Shire Publications.

———. 1988b. "Florists' Societies and Feasts after 1750." *Garden History* 11 (Spring): 11.

Genders, Roy. 1961. *Miniature Chrysanthemums and Koreans*. London: Blandford Press.

———. 1971. *Collecting Antique Plants: The History and Culture of the Old Florists' Flowers*. London: Pelham Books.

Haworth, A. H. 1833. "A New Arrangement of the Double-Flowered Chinese Chrysanthemums, with an Improved Method of Cultivation." *Gardener's Magazine* 9 (April): 218–26; reprinted in *Floricultural Cabinet*, June 1, 73–80.

Henderson, Peter. 1875. *Gardening for Pleasure*. New York: Orange Judd. Reissued 1888.

———. 1903. *Practical Horticulture*. New York: Orange Judd. Reissued 1911.

Herrington, Arthur. 1905. *The Chrysanthemum: Its Culture for Professional Growers and Amateurs*. New York: Orange Judd.

Hovey, Charles Mason. 1846. "Descriptive Account of Twenty-Four New Varieties of Chrysanthemums, with Some Observations on Their Cultivation." *Magazine of Horticulture* 12 (June): 213–16.

Lochot, J. 1930. *Le Chrysanthème*. 4th ed. Paris: Librairie Agricole de la Maison Rustique, pp. 178–97.

Loudon, J. C. 1822. *An Encyclopedia of Gardening*. S.v. "Chrysanthemum."

———. 1833. *Gardener's Magazine*.

Miller, Wilhelm. 1898. *Fourth Report upon Chrysanthemums*. Ithaca, NY: Cornell University.

Morton, James. 1891. *Chrysanthemum Culture for America*. New York: Rural Publishing.

Robinson, William. 1895. *Gardening World*.

———. 1900. *The Garden: An Illustrated Weekly Journal of Gardening* 60:585.

———. 1902. "Obituary of Simon Délaux." *The Garden: An Illustrated Weekly Journal of Gardening* 62 (July 5): 2.

———. 1902. "National Chrysanthemum." *The Garden: An Illustrated Weekly Journal of Gardening* 62 (December 6): 222.
Sabine, Joseph. 1821. "Account and Description of the Various Types of Chinese Chrysanthemums." *Horticultural Transactions* 4 p. 334, vol. 5, p. 12
Sargent, Charles Sprague, ed. 1893. "The New York Chrysanthemum Show." *Garden and Forest* 6:478.
Smith, Elmer D. 1904. *Smith's Chrysanthemum Manual.* Adrian, MI: privately published.
Spaargaren, J. J. 2016. *Chrysanthemum: Origin and Spreading.* Aalsmer, Netherlands: privately printed.
Transactions of the Massachusetts Horticultural Society. 1888.

Chapter 3: Penstemon

Farrer, Reginald. 1919. *The English Rock-Garden.* London: T. C. and E. C. Jack.
Floral Magazine. 1870. "Penstemons—Agnes Laing and Stanstead Surprise." Vol. 9, plate 469.
Lindgren, Dale T. 2006. *List and Description of Named Cultivars in the Genus Penstemon.* Lincoln: University of Nebraska–Lincoln Extension and the American Penstemon Society.
———. 2010. *APS Officers 1947 to 2010.* n.p.: American Penstemon Society.
Lindgren, Dale T., and Ellen Wilde. 2003. *Growing Penstemons: Species, Cultivars and Hybrids.* Haverford, PA: Infinity Publishing.
Linnaeus, Carolus. 1753. *Species Plantarum.* Holmia (Stockholm): Impensis Laurentii Salvii.
Mitchell, John. 1748. "Dissertatio brevis de principiis botanicorum et zoologorum deque novo stabiliendo naturæ rerum congruo." Appendix to *Acta physico-medica academiæ caesareæ Leopoldino-Franciscanae naturæ curiosorum exhibentia ephemerides sive observationes historias et experimenta.* Vol. 8, pp. 187–202. Nuremberg: B. W. M. Endteri Consortium.
Nold, Robert. 1999. *Penstemons.* Portland, OR: Timber Press.
Taylor, Judith M. 2014. *Visions of Loveliness: Great Flower Breeders of the Past.* Athens, OH: Swallow Press / Ohio University Press.
Thompson, William. 1851. *English Flower Garden: A Monthly Magazine of Hardy and Half-Hardy Plants* 1:161–63.
Way, David, and Peter James. 1998. *The Gardener's Guide to Growing Penstemons.* Portland, OR: Timber Press.

Chapter 4: Gladiolus

Beal, Alvin C. 1916. *Gladiolus Studies I.* Extension Bulletin 9. Ithaca: New York State College of Agriculture at Cornell University.
Gerard, John. 1597. *Herball; or, A General History of Plantes.* London.

Hottes, Alfred C. 1916a. *Gladiolus Studies II*. Extension Bulletin 10. Ithaca: New York State College of Agriculture at Cornell University.

———. 1916b. *Gladiolus Studies III*. Extension Bulletin 11. Ithaca: New York State College of Agriculture at Cornell University.

Kaiser, Bernd. 2009. *Wilhelm Pfitzer—Kunst- und Handelsgärtner, Stuttgart und Fellbach—Diplome, Ehrenpreise, Medaillen*. Fellbach, Germany: Self-published.

Linnaeus, Carolus. 1753. *Species Plantarum*. Holmia (Stockholm): Impensis Laurentii Salvii.

MacSelf, A. J. 1925. *Gladioli*. London: Thornton Butterworth.

Massachusetts Horticultural Society. 1880. *History of the Massachusetts Horticultural Society, 1829–1878*.

Metzler, Walter. 2010. "Biographie Max Leichtlin." *Blick in die Geschichte* 89 (December 23). Karlsruhe Stadtgeschichte.

Miller, Philip. 1731 (?). *The Gardener's Dictionary*. London.

Parkinson, John. 1629. *Paradisi in Sole*. London: Lownes and Young.

Pearce, Bruce. 1929. "A Gladiolus Wizard." *Simcoe Reformer*, n.d.

Randhawa, G. S., and A. Mukhopadhyay. 1986. *Floriculture in India*. Reprint, New Delhi: Allied Publisher, 2010.

Skidelsky, S. S. 1916. *The Tales of a Traveler*. New York: De La Mare Publishing.

Summy, Carol. 1997. "Horticulturist to Be Recognized." *Goshen News*, May 12.

Thurber, George. 1878. *Reports and Awards: U. S. Centennial Commission Group XXIX*. Vol. 20, p. 395. Philadelphia, PA: J. B. Lippincott.

Chapter 5: Dianthus

Bailey, L. H. 1914. *Standard Cyclopedia of Horticulture*. New York: Macmillan.

Cook, Everett Thomas, ed. 1903. *Carnations, Picotees*. Reprint, Carlisle, MA: Applewood Books, c. 2000.

Cook, T. H., James Douglas, and J. F. McLeod. 1911. *Carnations and Pinks*. Reprint, Carlisle, MA: Applewood Books, c. 2000.

Crozat, Stephane, Philippe Marchenay, and Laurence Berard. 2010. *Fleurs, fruits, legumes l'epopée Lyonnaise*, 104–5. Lyon, France: Editions Lyonnaises d'Art et d'Histoire.

Darwin, Charles. 1868; reprint 1905. *The Variation of Animals and Plants under Domestication*. London: John Murray.

———. 1999. *The Correspondence of Charles Darwin*. Volume 11: *1863*. Edited by Frederick Burkhardt and James Secord. Cambridge: Cambridge University Press.

First International Dianthus Register. 1974. London: Royal Horticultural Society.

Futhey, J. Smith, and Gilbert Cope. 1881. *History of Chester County, Pennsylvania*, vol. 2. Philadelphia: L. H. Everts.

Galbally, John, and Eileen Galbally. 1997. *Carnations and Pinks for Garden and Greenhouse.* Portland, OR: Timber Press.
Genders, Roy. 1962. *Garden Pinks.* London: John Gifford.
Gerard, John. 1597. *Herball; or, A General History of Plantes.* London.
Holley, Winfred Davis, and Ralph Baker. 1991. *Carnation Production.* 2nd ed. Dubuque, IA: Kendall Hunt.
Inwersen, Will. 1949. *The Dianthus: A Flower Monograph.* London: Collins.
Lamborn, Levi Leslie. 1901. *American Carnation Culture.* 4th ed. Alliance, OH: privately printed.
Padilla, Victoria. 1961. *Southern California Gardens: An Illustrated History.* Berkeley: Universtiy of California Press.
Parkinson, John. 1629. *Paradisi in Sole.* London: Lownes and Young.
Skidelsky, S. S. 1916. *The Tales of a Traveler: Reminiscences and Reflections from Twenty-Eight Years on the Road.* New York: De La Mare Publishing.
Turner, William. 1551. *A New Herball.* London.
USDA, National Agricultural Statistics Service. 2013. *Floriculture Crops 2012 Summary.* Available at https://usda.mannlib.cornell.edu.
Ward, Charles Willis. 1903. *The American Carnation: How to Grow It.* Reprint, Carlisle, MA: Applewood Books, c. 2000.
Weguelin, H. W. 1900. *Carnations and Picotees for Garden and Exhibition, with a Chapter Concerning Pinks.* London: George Newnes.
Weston, Thomas A. 1931. *Practical Carnation Culture.* New York: A. T. de la Mare.

Chapter 6: Clematis

Clematis: The Journal of the British Clematis Society. 1991– .
Coats, Alice M. 1964. *Garden Shrubs and Their Histories.* New York: Dutton.
Evison, Raymond. 1998. *The Gardener's Guide to Growing Clematis.* Portland, OR: Timber Press.
Fisk, Jim. 1964. *Success with Clematis.* London: Thomas Nelson and Sons.
———. 1994. *Clematis: The Queen of Climbers.* London: Cassell.
Gardener's Chronicle. 1888. "Obituary: Thomas Cripps." April 21, p. 504.
Gerard, John. 1597. *Herball; or, A General History of Plantes.* London.
Gilbert, Edward James. 2012. *The Cripps Nursery.*
Johnson, Magnus. 2001. *The Genus Clematis.* Södertalje, Sweden: privately published.
Lloyd, Christopher, with Tom Bennett. 1979. *Clematis.* London: Viking Press.
Marczyński, Szczepan. 2014. "Wladyslaw Noll." *Clematis International,* 82.
Markham, Ernest. 1935. *Clematis.* London: Country Life.
Tropicos. 2017. Source for *C. verticillaris* DC, *C. douglasii* Hook., *C. quinquefoliolata* Hutch. (not in GRIN). Searchable database available at www.tropicos.org. St. Louis: Missouri Botanical Garden.

Willson, Eleanor Joan. 1989. *Nurserymen to the World*. London: privately published.
Woolfenden, Ken. 2010. "Memorial to Brother Stefan Franczak." *Clematis International*, 27.

Chapter 7: Pansy/Viola

Brett, Walter. 1926. *Pansies, Violas and Violets*. London: Newnes.
Clayton Payne, Adam. 1989. *Flower Gardens of Victorian England*. New York: Rizzoli.
Cook, E. T. ed. 1903. *Sweet Violets and Pansies and Violets from Mountain and Plain*. Reprint, Whitefish, MT: Kessinger Publishing, 2010.
Crane, Daniel Burton. 1912. *Violas, Pansies and Violets*. London: Collingridge.
Cuthbertson, William. 1910. *Pansies, Violas and Violas*. London: T. C. and E. C. Jack.
Darwin, Charles. 1868. *The Variation of Animals and Plants under Domestication*. London: John Murray.
Desmond, Ray. 1980. *Loudon and Nineteenth Century Horticultural Journalism*. Washington, DC: Dumbarton Oaks, Trustees of Harvard University.
———. 1987. *A Celebration of Flowers: Two Hundred Years of* Curtis's Botanical Magazine. London: Twickenham Royal Botanic Gardens at Kew with Collingridge Publishers.
Duthie, Ruth. 1984. *English Florists' Flowers and Societies*. Aylesbury, UK: Bucks Shire Publications.
Elliott, Brent. 1986. *Victorian Gardens*. Portland, OR: Timber Press.
Farrar, Elizabeth. 1989. *Pansies, Violas and Sweet Violets*. Hurst, Berkshire, UK: Hurst Village Publishing.
Fuller, Rodney. 1990. *Pansies, Violas and Violettas*. Ramsbury, Wiltshire, UK: Crowood Press.
Gäde, Helmut. 2010. *Wege und Umwege* [Roads and detours]. Quedlinburg, Germany: Ein Stadtführer—Exkurs.
Genders, Roy. 1958. *Pansies, Violas and Violets*. London: John Gifford.
Henderson, George J. 1885. In *Glenny's Garden Almanach*.
Jordan, Charles, Jack Ballantyne, Jessie K. Burnie, and William Cuthbertson. 1898. *Pansies, Violas, and Violets*. London: Macmillan.
M'Intosh, Charles. 1844. *The Flower Garden*. London: William S. Orr.
———. 1868. *The Book of the Garden*. London: William Blackwood and Sons.
Newcomb, Peggy Cornett. 1985. *Popular Annuals of Eastern North America, 1865 to 1914*. Cambridge, MA: Harvard University Press.
Scott-James, Anne, Ray Desmond, and Frances Wood. 1989. *The British Museum Book of Flowers*. London: British Museum.

Thomson, William. 1841. "History of the Heartsease." *Floricultural Cabinet* 9:222–25.
Wittrock, V. B. 1895. "Viola-Studier II." *Acta Horti Bergiani* 2, no. 7.

Chapter 8: Water Lily

Bisset, Peter. 1909. *The Book of Water Gardening.* New York: A. T. de la Mare Printing and Publishing.

Conard, Henry S. 1905. *The Water-Lilies: A Monograph of the Genus Nymphaea.* Washington, DC: Carnegie Institution of Washington.

Knott, Kit. 2006. Our Adventure with *Victoria.* www.victoria-adventure.org.

Masters, Charles Otto. 1974. *Encyclopedia of the Water-lily.* Neptune City, NJ: TFH Publications.

Slocum, Perry D., and Peter Robinson, with Frances Perry. 1996. *Water Gardening: Water Lilies and Lotuses.* Portland, OR: Timber Press.

Tricker, William. 1897. *The Water Garden.* Reprint, Whitefish, MT: Kessinger Publishing, 2010.

INDEX

Page numbers in **boldface** refer to illustrations.

Albert, Charles, 181
Alegatière, Alphonse, 119, 120, 121, 130–31
Allard, Henry, 13
Allen, C. L., 53, 100
Allwood, Montague, and brothers, 123–25, 127, 130
American Carnation Society, 132, 141
American Gladiolus Society, 99, 107
Amherst, Sarah, Countess, 144
Audiguier, Dr., 39, 50
Austin, Iva, 98–99
Auten, B. C., 99
Azalealand, 14

Bachman, John, 9
Bacon, Francis, 189
Baerman, Ralph, 111
Bailey, Liberty Hyde, 67
Baltet, Charles, 39, 154
Barr, William, 60
Bartram, John, 8
Baur, Adolphe, 131–32
Baur, Francis, 132
Baur & Steinkamp Company, 15, 132
Beal, Alvin C., 86–87, **87**, 95–97, 101
Bedinghaus, Hermann Josef, 84–85, 98
begonias, 1, 2, 75, 204
Benary, Ernst, 126, 174, 184
Bennet, Emma, 177
Bennet, Mary Elizabeth, 170, 177–78
Bergianus Botanic Garden, 167
Bernard, M., 39, 50
Bernet, Marc, 36, 39–40, **40**, 45, 47
Beskaravajnaja, Margarita Alexeevna, **162**, 162–63, 164
Beutler, Linda, 165
Bidwell, John, 86, 138–39
Bisset, Peter, 194
Blancard, Pierre-Louis, 19, 21, 23, **41**, 41–42
Blick, Charles, 32
Bockmann, Herr, 183
Boisselot, Auguste, 154
Bolton, Robert, 123
Bonamy et Frères, 42
Bonnefous, A., 42
Bonpland, Aimé, 193
Boucharlat, Laurent, 42–43, 48

Bouché, Claude, 192
Bourdet, Louis, 14
Bowles, E. A., 195
Bridges, Mr., 194
Brown, E. and C., 128, 169
Brown, Thomas, 50, 73, 87–88
Buccleuch Nurseries, 72–73
Bugnot, Jules, 174, 181
Buist, Robert, 8–9, 137
Bull, William, Sr., 30, 33
Burbank, Luther, 3, 89, 97–98, 99
Burbidge, F. W., 32
Burchett, George, 99–100
Burnett, H., 130
Burpee, W. Atlee, 88, 187
Butt, Leonard, 111

Calderon de la Barca, Fanny, 11
Calvat, Ernest, 43, 44, 47
Cannell, Henry, 30
Carey, Major, 36, 37
carnations and pinks, 1, 113–41; classification of, 114–15; dissemination of, 113, 117, 121–22; nomenclature for, 113–14
   CULTIVARS: 'Adonis', 141; 'Arc en Ciel', 131; 'Armazindy', 132; 'Astoria', 121; 'Atim', 119, 131; 'Aviator', 133; 'Beauty of Healey', **118**; 'Bidwell', 138; 'Bielson' (aka 'Biohon'), 117, 131; 'Buttercup', 120; 'Chester Pride', 120; 'Crusader', 132; 'Doris', **124**, 125; 'Edwardsii', 119, 121, 131; 'Eldorado', 139; 'Emma Wocher', 141; 'Enchantress', 127, 136; 'Estelle', 141; 'Eulalie', 139; 'Evelina', 141; 'Fiancée', 121, 133–34; 'Flatbush', 120; 'Golden Gate', 139; 'Harlowarden', 133; 'Indiana Markey', 132; 'Indianapolis', 132; 'John Ball', **118**; 'Kitty Clever', 139; 'Lady Emma', 120, 121; 'Laura Jane', **118**; 'La Pureté', 119, 120, 121, 131; 'Mahon', 117, 131; Malmaison series, 120; 'Manteaux Royal', 120; 'Miss Eaton', **118**; 'Miss Nightingale', **118**; 'Mont Blanc', 120; 'Mrs. Degraw', 120; 'Mrs. H. Burnett', 130; 'Mrs. Lawson', 136; 'Mrs. N. M. Higinbotham', 132; 'Mrs. Sinkins', 122, **129**; 'Mrs. W. T. Lawson', 130; 'Old Velvet', **118**; 'Pacific', 139; 'Paradise', 139; 'Piru', 139; 'Pride of Market', 130; 'Ramona', 139

carnations and pinks (*cont.*)
  HYBRIDS: *Dianthus* × *allwoodii* series, 124–25; *D. caryophyllus* × *barbatus* ("Fairchild's mule"), 113, 116, **117**
  SPECIES: *Dianthus barbatus*, 113, 115, 130; *D. caryophyllus*, 113, **114**, 115, 125; *D. chinensis*, 115; *D. plumarius*, 115, 125; *D. repens*, 114
Carr, Robert, 8, 11
Carré, Alfred, 154
Cassier, M., 174, 182, 183, 186
Chandler's Nursery, 24
Chantrier, Alfred, 43
Charmet, André, 44, 50
Charpentier, M., 182
Chelsea Physic Garden, 23, 24, 82
Chicago Carnation Company, 132–33
Childs, John Lewis, 53, 57, 98, 100, 108
Christen, Louis, 155
chrysanthemums, 19–63; classification of, 21, 27; dissemination of, 19, 21–23, 36; France vs. England, 38; nomenclature for, 21–22, 45
  CULTIVARS: 'Ada Spaulding', 59; 'Addie Decker', 59; 'Advance', 63; 'Alaska', 62; 'Alfred Salter', 33; 'Alfred Warne', 62; 'Alice Brown', 60; 'Amaterasu', 61; 'Ami Jules Chrétien', 44; 'Annibal', 39; 'Annie Salter', 24, 33; 'Apricot Courtier', **28**; 'Arizona', 59; 'Aureole Virginale', 32; 'Baronne de Staël', 40; 'Beaumont', 37; 'Beauty of Kingsessing', 62; 'Belle Hickey', 55; 'Ben d'Or', 36, 56; 'Bessie Godfrey', 32; 'Bob', 25; 'Bohemia', 59; 'Bouquet Fait', 56, 57; 'Bouquet Nationale', 56, 57; 'Bridesmaid', 32; 'Brooklyn', 54; 'Brynwood', 59; 'Bryant', 55; 'Cambridge', 62; 'Candeur des Pyrenées', 43; 'C. A. Reeser', 57; 'Carry Denny', 57; 'Catros Gerarde', 44; 'Changeable Buff', 51; 'Charles Blick', 32; 'Chatillon', 46; 'Christmas Eve', 56; 'Cisco', **28**; 'Clara Barton', 54; 'Clarence', 55; 'Cloth of Gold', 54; 'Colossal', 60; 'Columbia', 54; 'Commotion', 59; 'Comte de Germiny', 36; 'Comte F. Lurani', 44; 'Connecticut', 59; 'Coquette de Chatillon', 46; 'Cornell', 54; 'Coronet', 60; 'Count of Germany', 56; 'Crimson', 51; 'Cullingford', 56; 'Curled Lilac', 51; 'Cyclone', 59; 'The Czar' (aka 'Peter the Great'), 37; 'Dango Zaka', 60; 'Delightful', 32; 'Delistar', **28**; 'Duc d'Albuféra', 39; 'Duchess', 62; 'Duchess of Connaught', 36; 'Duchesse d'Orléans,' 43; 'Duchess of Devonshire', 32; 'Edwin Booth', 60; 'Eglantine', 58; 'Elaine', 37, 56; 'Eleanor Oakley', 59; 'Elkshorn', 62; 'Emily Selinger', 55; 'Emmie Ricker', 55; 'Ernest Fierens', 33; 'E. S. Renwick', 59; 'Eucharis', 42; 'Eugene Dailledouze', 53; 'Excellent', 63; 'Exmouth Crimson', 32; 'Exmouth Rival', 32; 'Fabias de Mediana', 44; 'Fair Maid of Guernsey', 37, 53, 56; 'Fannie Block', 60; 'Fantasie', 56; 'Flocon de Neige', 44; 'Flora', 55; 'Frank Wilcox', 60; 'F. T. McFadden', 60; 'Garnet', 59; 'George Atkinson', 59; 'George McClure', 59; 'Georges Sand', 39–40; 'G. F. Moseman', 59; 'Gladys Spaulding', 59; 'Gloria Rayonnante', 50; 'Gloriosum', 56, 62; 'Gold Coast', 33; 'Gold Bordered Red', 27; 'Golden Ball', 54; 'Golden Dragon', 56; 'Golden Gate', **28**; 'Golden Lotus', 51; 'Golden Yellow', 53; 'Goliath', 37; 'G. P. Rawson', 60; 'Grand Napoleon', 39; 'Hackney Homes', 37; 'H. A. Gane', 55; 'Henry Ward Beecher', 54; 'Hilda Tilch', 32; 'Hobson', 54; 'H. Waterer', 62; 'Indian', 59; 'Iona', 59; 'Iowa', 59; 'Iroquois', 59; 'Ithaca', 59; 'Ivory', 53, 63; 'James F. Mann', 55; 'James R. Pitcher', 59; 'James Salter', 56; 'Japonais', 44; 'J. Collins', 56, 62; 'Jean Humphreys', 59; 'Jeanne d'Arc', 44; 'John Lane', 57; 'John M. Hughes', 62; 'Junior Webster', 55; 'Juno', 59; 'Kansas', 59; 'KelvinMandarin', **28**; 'The Khedive', 37; 'Kildare', 58; 'Kimie', **28**; 'Kioto', 55; 'Lady Carey', 37; 'Lady Slade', 56; 'Large Lilac', 51; 'La Triomphante', 48; 'L. Canning', 62; 'Leopard', 60; 'Lilac', 53; 'Lilian B. Bird', 55; 'Lizzie Cartledge', 32; 'Lizzie Gannon', 55; 'Lone Star', **28**; 'Lord Byron', 56; 'Lord Rosebery', 33; 'Lucrece', 62; 'Magicienne', 33; 'Magnet', 62; 'Maid of Athens', 56; 'Maréchal Maison', 39; 'Marian', 55; 'Maria Ward', 59; 'Market Favourite', 32; 'M. Astory', 44; 'Medee', 42; 'Minnewawa', 59; 'Miss Anna Hartshorne', 62; 'Miss C. Harris', 62; 'Miss Dorothy Shea', 32; 'Miss Mary Wheeler', 63; 'Miss Meredith', 62; 'Miss Sue Waldron', 60; 'Mlle. Lacroix', 44; 'Mlle. Marthe', 25; 'Mme Berthier Rendatler', 44; 'Mme Custex Desgranes', 42; 'Mme Grame', 56; 'Mode', 57; 'Model of Perfection', 25; 'Mohawk,' 59; 'Monadnock', 62; 'Monsieur Chas Molin', 32; 'Mont Blanc', 62; 'Moonflower', 60; 'Moonlight', 56; 'Mountain of Snow', 63; 'Mrs. A. Blanck', 62; 'Mrs. A. C. Burpy', 63; 'Mrs. Alpheus Hardy', 32, 51, **52**, 55, 59; 'Mrs. Andrew Carnegie', 61; 'Mrs. Anthony Waterer', 62; 'Mrs. Brett', 56; 'Mrs. C. Carey', 37; 'Mrs. C. H. Wheeler', 56, 62; 'Mrs. C. L. Allen', 56, 57; 'Mrs. Cleveland', 60; 'Mrs. Cornelius Vanderbilt', 59; 'Mrs. DeWitt Smith', 59; 'Mrs. E. Miles', 38; 'Mrs. Fottler', 55; 'Mrs. George Bullock', 62; 'Mrs. G. W. Coleman', 62; 'Mrs. Haliburton', 38; 'Mrs. Henry Evans', 61; 'Mrs. Huffington', 37; 'Mrs. Irving Clark', 63; 'Mrs. J. N. Gerard', 60; 'Mrs. Joel J. Bailey', 62; 'Mrs. John N. May', 63; 'Mrs. John Pettit', 59;

'Mrs. John Wanamaker', 62; 'Mrs. Judge Benedict', 59; 'Mrs. Langtry', 60; 'Mrs. M. J. Thomas', 63; 'Mrs. Pethers', 37; 'Mrs. Potter', 60; 'Mrs. R. Mason', 62; 'Mrs. Sam Houston', 62; 'Mrs. S. Coleman', 59; Mrs. S. Lyon', 56;'Mrs. T. C. Price', 63; 'Mrs. Thomas A. Edison', 59; 'Mrs. T. H. Spaulding', 60; 'Mrs. Vannaman', 62; 'Mrs. Wanamaker', 59; 'Mrs. William Barr', 60; 'Mrs. William Howell', 62; 'Mrs. Winthrop Sargent', 57; 'Mrs. W. J. Godfrey', 32; 'Mrs. W. K. Harris', 63; 'Mt. Shasta', **28**; 'Nahanton', 55; 'Neesima', 55; 'Nemasket', 31, 58; 'Nevada', 55; 'Nil Desperandum', 62; 'Nippon Medusa', 55; 'Nonpareil', 32; 'Norfolk Hero', 32; 'Old Purple', **20**, 23, 42; 'Owen's Perfection', 33; 'Pagoda', 60; 'Paper White', 51; 'Parasol', 44; 'Parks', 53; 'Passiac', 59; 'Patricia Grace', 61; 'Pauline', 60; 'Peacock', **28**; 'Peculiarity', 60; 'Peggy Stevens', **28**; 'Pequot', 59; 'Perle Chatillonnaise', 46; 'Pink', 53; 'Pontiac', 62; 'Prefet Robert', 35; 'President Arthur', 62; 'President Hyde', 55; 'Président Loubert', 46; 'President Smith', 53; 'Président Truffaut', 46; 'Pride of Maidenhead', 33; 'Prince Alfred', 37; 'Prince Kamoutska', 60; 'Prince of Wales', 37; 'Princesse Pauline', 40; 'Princess of Wales', 33, 37; 'Public Ledger', 62; 'Puritan', 59, 62; 'Queen of England', 24; 'Quilled Flame', 51; 'Quilled Lilac', 53; 'Quilled White', 53; 'Ramona', 62; 'Raymond Poincaré', 46; 'Red Dragon', 56; 'Red Gauntlet', 37; 'Reine Blanche', 39; 'R. E. Jennings', 59; 'Reward', 57; 'Robert Craig', 63; 'Robert Crawford', 62; 'Robert Petfield', 33; 'Raleigh', 59; 'Rohallion', 58–59;'Rose Croix', 39; 'R. Walcott', 62; 'Sadie Martinot', 60; 'Sarah Bernhardt', 46; 'Sarnia Glory', 37; 'Savannah', 62; 'S. B. Dana', 55; 'Seatons Ruby', **28**; 'Semi-quilled White', 53; 'Semiramis', 62; 'Senkyo Kenshin', **28**; 'Sensation', 32; 'Shamrock', 54; 'Shasta', 62; 'SirIsaac Brock', 37; 'Sir Stafford Carey', 37; 'Small Yellow', 53; 'Snowdrift', 60; 'Snowstorm', 62; 'Sokoto', 60; 'Soleil Levant', 50; 'Sonce d'Or', 56, 57; 'Sport', 62; 'Stars and Stripes', 62; 'St Tropez', **28**; 'Sunnyside', 63; 'Sunset', 60; 'Tacomah', 62; 'Takaki', 59; 'Talford Salter', 56; 'Tassled White', 51; 'T. F. Martin', 60; 'Thisbe', 42;'Thomas Cartledge', 62;'Thunberg', 36; 'Tusaka', 59; 'Twilight', 57; 'V. H. Hallock', 60; 'Victoria', 37; 'Violet Rose', 63; 'Virginia', 59; 'Viviand Morel', 45; 'Wenonah', 62; 'We Wa', 59; 'W. Falconer', 60; 'Whirlwind', 60; 'White', 53; 'White Cap', 57; 'White Cheifton', 61; 'White Maud Dean', 53; 'White Treveana', 55; 'William Dewar', 62; 'William H. Lincoln', 55; 'William Turner', 46; 'W. J. Bryan', 54; 'Wonderful', 62; 'W. W. Coles', 63; 'Yellow Boy', 32; 'Yellow Eagle', 56; 'Yokohama Orange', 37; 'Zenobie', 59
HYBRIDS: 'William Penn', 51, 58
SPECIES: *Chrysanthemum indicum*, 21, 24, 25; *C. morifolium*, 19–21 (**20**), 23, 25; Chusan Daisy, 24, 25; *Dendranthema grandiflorum*, 21; *D. indicum*, 21; *D. japonicum*, 21
Chrysanthemum Society of America, 21, 61, 132
Chrysanthemum Society of New York, 53
Clark, James, 199
clematis, 143–65; dissemination of, 143–45; Japanese tea ceremonies and, 157–58
CULTIVARS: 'Aliosha', 163; 'Annie Wood', 146; 'Asao', 157; 'Captivation', 146; 'Comtesse de Bouchard', 155; 'Eugene Delattre', 155; 'General Sikorski', 161; 'Gloire de St Julien', 154 'Gloria Mundi', 146; 'Gravetye Manor', 149; 'Guiding Star', 146; 'Hendersonii', 147–49 (**148**); 'Jackmanii', 145, 149; 'Jackmanii Superba', 146; 'Jadwiga Teresa', 161; 'James Mason', 147; 'John Paul II', 160; 'Kakio', 157; 'Lady Caroline Nevill', 146; 'Madame Boselli', 155; 'Madame Edouard Andre', 155; 'Marcel Moser', 157; 'Mme Van Houte', 146; 'Mrs Markham', 149; 'Nellie Moser', 157; 'Nikita', **162**; 'Niobe', **161**; 'Papa Christen', 155; 'Stolen Kiss', 149; 'Uno Kivistik', 154; 'Victoria', 146; 'Ville de Lyon', 155; 'Viviany Morel', 155; 'Vyvyan Pennell', 153
HYBRIDS: *Clematis × eriostemon*, 147; *C. pitcheri × C. coccinea*, 157
SPECIES: *Clematis addisonii*, 145; *C. aethusifolia*, 144; *C. afoliata*, 145; *C. alpina*, 144; *C. armandii*, 145;*C. campaniflora*, 144; *C. chrysocoma*, 145; *C. cirrhosa*, 144; *C. coccinea*, 155; *C. crispa*, 145; *C. douglasii*, 145; *C. fargesii*, 145; *C. flammula* L., 144; *C. florida*, 144; *C. fusca*, 144; *C. grata*, 144; *C. heracleifolia*, 144; *C. integrifolia*, 144, 148, 157; *C. lanuginosa*, 144, 145, 149; *C. macropetala*, 144; *C. meyeniana*, 144; *C. montana*, 144; *C. ochroleuca*, 145; *C. orientalis*, 145; *C. paniculata*, 144–45; *C. patens*, 144, 145; *C. quinquefoliolata*, 145; *C. recta*, 144; *C. serratifolia*, 145; *C. songarica*, 145; *C. spooneri*, 145; *C. stans*, 144; *C. tangutica*, 145; *C. verticillaris*, 145; *C. viorna*, 144; *C. virginiana*, 145; *C. vitalba*, 143, 144; *C. viticella*, 143–44, 148
Clibran and Sons, W., 31
Coindre, Pierre, 47
Colvill and Son, James, 23, 83–84, 93
Condon, John, 53–54
Connard, Henry S., 200
Conrad & Jones Company, 174, 183

Cooper, Madison, 88
Cope, Caleb, 194
copyright and patent issues, 11–12, 40, 201
Cordonnier, Anatole, 43
Cottage Gardens Company, 120, 134
Cowee, Arthur, 88, 90, 100–101, **101–3**, 103
Craft, George, 96–97
Craig, Robert, 62–63
Crane, D. B., 175
Creighton, George, 134
Cripps, Thomas and Ellen, 145–46
Crozy, Michel, 50
Crozy, Pierre, 50
Curtis, William, 82–83
Curtis & Cobb, 97
Cuthbertson, William, 176

dahlias, 16, 22, 77, 109, 111, 123, 129, 164
Dailledouze and Zeller, 120, 134, **135**, 138
daisies, 3, 25
Daisy Hill Nursery, 195
Dalmais, M., 117, 119, 131
Darwin, Charles, 129–30, 177
Davenport, Nathaniel, 53
Davies, Ray, 194
Davis, James, 36–37
Dawnton, James, 36–37
Dean, Richard, 178, 179
Dean, William, 170, **171**, 172, 178
Decorah Gladiolus Gardens, 104
De Goede, Simon, 94
Délaux, Simon, 38, 39, 44
De May, M., 182
deutzias, 2, 97
Devonshire, William Cavendish, Duke of, 69, 91, 192
dianthus. See carnations and pinks
Dickson Nursery, 169, 172, 179
Dobbies Nursery, 175, 176
Doeppler, Herr, 184
Dominy, John, 35
d'Orbigny, Alcide, 193
Dorner, Frederick, 54, 121, 133, 134, **135**
Douglas, David, 67
Douglas, James, 125–26
Downie, John, 172
Downie, Laird, & Laing, 69, 172
Dreer, Henry, 198, 205
Drewlow, Lyndon, 14
Drummond, James, 69
Drummond, **97**; William, 69
Duchet, Joseph, 131
Dutch Horticultural Society, 111
Duval, Jacques, 182

Ecke, Albert, and family, 12–13, **13**, 14, 16
Edenside Nursery, 126–27
Eichholz, Henry, 136
Elizabeth I, 143–44
Ellwanger, Georg, 54

Emmons, Robert L., 53
Engelmann, Carl G., 127–28, 175
English National Chrysanthemum Society, 21
Enteman, Mrs., 14
Exmouth Nurseries, 32

Fairchild, Thomas, 113, 116
Farrer, Reginald, 69
Fate, Fred, 79
Feder, Henry, 139
Fewkes & Sons, Edwin, 54–55
Fischer, Carl, 111
Fisher, A. Sewall, 136
Fisher, Peter, 127, 136
Fisk, Jim, 146, 160, 161
Flanagan & Nutting, 67
Fleming, John, 171
Forbes, John, 2–3, 25, 72–74, 75
Forster, Robert, 31
Fortune, Robert, 25, 144, 152
Fox, Francis, 83
Franczak, Stefan, 147, 158–60 (**159**), 161
Fraser, John, 67
Freestone, John, 24, 32
Fretwell, Barry, 147
Froebel, Otto, 197, 198
Fry, Joel, 10
Furcraea, 84–85

Gäde, Helmut, 186
Galleotti, Henri, 69
Gambier, James, Lord, 169
Garner, Wrightman, 13
Gebrüder Mette. See Mette Brothers
Genders, Roy, 24, 25, 175, 176
Gerard, J. N., 60
Gerard, John, 81, 115, 143
Gérard, René, 157
gladioli, 2, 3, 81–111; nomenclature for, 81, 86
  CULTIVARS: 'Amanda Mahy', 93; 'American', 107; 'Baron J. Hulot', 108; 'Black Beauty', 107; 'Blushing Bride', 93, 111; 'The Bride', 93; 'Cardinal', 107; 'Ceres', 95; Childsii series, 98, 108; 'Colonel Wilder Wright', 96; 'Elnora', 96; 'Empress of India', 107; 'Eugene Scribe', 95; 'Evelyn Kirkland', 107; 'Exemplar', 96; 'Fairy', 107; 'Fire King', 107; 'Florence', 107; 'Fusilier', 93; 'Guernsey Glory', 93; 'Hauptmann Kohl', 109; 'Innocence', **97**; 'Jeanie Dean', 96; 'Jenny Lind', 104; Kelwayi varieties, 91; 'Kunderdi Glory', 106; 'Laciniatus', 106; Leichtilini varieties, 108; 'Miranda', 93; 'Mrs. Calvin Coolidge', 107; 'Neues Jahrhundert', **110**; 'Nymph', 111; 'Panama', 107; 'Peer Gynt', 104; 'Picardy', 111; 'Pure Bride', 93; 'Reinhold Breitschwert', **110**; 'Rita Page', 93; 'Salmonia', 96; 'Snowbank', 103; 'Snow Princess', 109; 'Spitfire', 111; 'Sulphur King', 107; 'Victory', 103

HYBRIDS: *Gladiolus colvilli*, 84; *G.* × *gandavensis*, 84–85, 95, 109, 111; *G. ramosus*, 85, 95, 111; Langprim series, 92
SPECIES: *Gladiolus alatus*, 82; *G. aurantiacus*, 86; *G. blandus*, 82, 86; *G. brenchleyensis*, 111; *G. byzantinus*, 81, 86; *G. cardinalis*, 82–83, 86, 93, 111; *G. communis*, 81; *G. cruentus*, 86, 108; *G. dalenii*, 85, 86; *G. dracocephalus*, 86, 94–95; *G. floribundus*, 95; *G. primulinus*, **82**, 83, 86, 92, 93; *G. psittacina*, 83; *G. psittacinus*, 86; *G. purpuroaureatus*, 94; *G. recuirvus*, 82; *G. trimaculatus*, 86; *G. tristis*, 82, 84, 86, 95; *G. venustus* 93
Gladiolus Breeders' Association, 93
Gladiolus Hall of Fame, 106
Godefroy Lebeuf, 108
Godfrey, W. J., 32
Gotthold & Company, 184
Graham, Robert, 10
Grasshoff, Martin, 185
Gray, Asa, 67
Grieve, James, 169, 172–74 (**173**), 178–80
Groff, Henry Harris, 3, 33, 88–90 (**89**), 98, 100–101, 104, 131
Gurney, James, 195
Gutbier, Gregor, 6, 17

Haenke, Thaddäus, 193
Hallock & Son, V. H., 60, 98, 108
Hamilton, William, 197
Hardy, Graham, 179
Hardy Plant Farm, 195
Harrison, George, 24
Hartshorn, James, 121, 132–34
Hawick Horticultural Society, 73
Haworth, A. H., 23, 27
Hegg, Thormod, 16–17
Hemus, Hilda, 99
Henderson, Alfred, 138
Henderson, Andrew, 172
Henderson, Peter, 55–57 (**56**), 136–38, 187
Heraud, Jean, 50
Herbemont, Mrs. Nicholas, 11
Herbert, William, 83, 85–86
Herrera, Alonso de, 122
Higinbotham, James, 132
Hill, Edward Gurney, 57
Hoeg, Christian, 104–5
Hollis, George, 58
Honn, C. E., 54
Hooker, Joseph, 86
Horn, Maurice, 158
Horticultural Society of London, 23–25. See also Royal Horticultural Society
horticulture journals, 24, 27, 31, 88, 99
Hosp, Franz, 122
Hoste, M., 39, 44, 50
Hottes, Alfred, 86–88, 91
Hovey, Charles Mason, 51
Hovey, William, 96

International Union for the Protection of New Varieties of Plants, 11

Jackman, George, and sons, 146, 147, 149
James, Peter, 69
Jekyll, Gertrude, 169
Johnson, Anna, 71
Johnson, Magnus, 147, 152, 155, 164–65, **165**
Joliet Floral Company, 132

Kaiser, Bernd, 109
Keating, William, 11
Kelway, James, and sons, 90–92, 93
Keynes, John, 123, 147
Kilvington, Robert, 51, 58
Kingman, C. D., 58
Kivistik, Uno and Aili, **153**, 153–54
Klager, Hulda, 99
Klatt, Wilhelm, 86
Knappwer, A., 184
Knott, Kit, 196
Koch, Karl Heinrich, 84
Krelage family, 109, 111
Kunderd, Amos E., 99, 106–7

Lachot, J., 25
Lack, H. Walter, 9
Lacroix, Etienne, 39, 44
Lacroix, Louis, 38, 39, 45
Lamarck, Jean-Baptiste, 86, 163
Lamborn, Levi, 120–21, 133
Landreth, David, 96
Lane, John, 58
Layn, Jacob, 23
Lebois, Emile, 36, 40, 45
Lee, James, 170
Leichtlin, Max, 98, 108
Lemaire, Louis-Jules, 45
Lemaire, Paulette, 45
Lemoine, Victor, 2–3, 47; chrysanthemums and, 38–39, 46; clematis and, 155; gladioli and, 94–95, 99, 108; lilacs and, 46, 74; penstemons and, 2, 69, 72, 74–75
Lenton, Stephen, 122, 138–39
Liecht family, 200
lilacs, 46, 74, 99
Lilypons Water Gardens, 200, 204
Lindgren, Dale, 73, 78–79
Lingg, Joseph, 200
Linnean Gardens, 95
Linnaeus, Carolus, 24, 66–67, 81, 113
Lorenz, C., 184
lotuses. See water lilies
Loudon, J. C., 24, 27
Low, Simon, and Co., 128
Luckie, G. J., 194
Lysenko, Trofim, 163

Maclure, William, 10
Mahy family, 93–94

Marc, Charles, 120
Marczyski, Szczepan, 160–61
Margottin, Jacques Julien, 182
Markham, Ernest, 149, **150–51**
Marliac, Joseph Latour-, 195–98 (**196**), 203
Marrouch, M., 45
Massachusetts Horticultural Society, 51, 55, 58, 62, 96
McNabb, James, 9
McTear, James, 96–97
Mendel, Gregor, 98, 121
Mette, Peter Heinrich, **185**
Mette Brothers, 184–86, 187
Miellez, Auguste, 40, 46, 172, 182
Mikkelsens Inc., 14, 16
Miller, George W., 58
Miller, Philip, 24, 82
M'Intosh, Charles, 170
Missouri Botanical Garden, 201
Mitchell, Eugene H., 51
Mitchell, John, 65–66
M'Mahon, Bernard, 95
Morel, Francisque, 155–56
Morgan, Hugh, 143–44
Moschkowitz und Siegling, 186
Moser, M., 157
Morton, James, 36, 38, 51, 53, 54–55, 57–60, 62

National Auricula and Primula Society, 126
National Carnation and Picotee Society, 117, 126, 129
National Chrysanthemum Society, 24, **28**, 31, 33, 58, 178
Nelson, Rolf and Anita, 200
New York Horticultural Society, 61
Noble, Charles, 151–52
Noisette, Louis, 24–25
Noll, Wladislaw August, 160–61
Nonin, Auguste, 46–47
Nonin, Henri, 47
North American Gladiolus Council, 86

Odier, James, 174, 182, 186
Oldfield Nurseries, 31
Oliver, William, 73
orchids, 1, 22; hybrid, 36; *Paphiopetalum tankervillae*, 177; Pring and, 201
Orlov, Mikhail, 162, 163
Ortgies, Eduard, 192
Osbeck, Pehr, 24
Owen, Robert, 30, 33
Ozawa, Kazushige, 157–58

Padilla, Victoria, 121–22
Palmer, E. F., 111
pansies and violets, 167–87; Belgian ("fancy"), 170, **171**, 172; dissemination of, 167, 170; English pansy societies, 174; nomenclature for, 167, 170, 172, 174, 179, 183

CULTIVARS: 'Azurea', 187; 'Blue Bell', 178; 'Blue King', 180; 'Bridesmaid', 178; 'Burpee's Defiance Pansies', 187; 'Diana', 175; 'Dr. Faust', 187; 'Eileen', 175; 'Engelmann's Giant', 175; 'Etoile du Nord', 172; 'Goldorange', 184; 'Henderson's Mammoth Butterfly Pansies', 187; 'Imperatrice Eugenie', 182; 'Kaiser Wilhelm', 184; 'Lady Gambier', 170; 'Leonidas', 182; 'Madora', 170; Masterpiece strain, 176; 'Napoleon III', 182; 'Negurfurst', 187; 'Ottile von Menzingen', 184; 'Princess Alice', 172; 'Princess Beatrice', 178; Princess violas, 176; Swiss Giant varieties, 176; 'Triumph of the Giants', 186; 'True Blue', 178; 'Victoria', 170; 'Winifred Philips', 175
HYBRIDS: F1 hybrids, 174, 176; *Viola* × *wittrockiana*, 167–68, **168**
SPECIES: *Viola altaica*, 168, 169, 183; *V. cornuta*, 179, 180; *V. lutea*, 168, 179, 184; *V. odorata*, 172; *V. striata*, 179; *V. tricolor*, 167–68, 183, 186
Parkinson, John, 115
Parkman, Francis, 97
Paul, William, 172
Paxton, Joseph, 69, 192, 194
pelargoniums, 2, 30; "Zonale," 75
Pélé, André-Philippe, 40, 45, 47
Pellier, Alfred, 69, 75
Pennell, Richard, 152
Pennell, Walter Everitt, 152–53
Pennsylvania Horticultural Society, 11, 51, 58, 59, 62–63
penstemons, 2, 3, 65–79 (**66**); classification of, 71; nomenclature for, 65–67
CULTIVARS: 'Crimson Gem', 73; 'Florist Giant', 73; 'Newbury Gem', 73; 'Newbury White Gem', 73; 'Pershore Pink Necklace', **68**; 'Prairie Dawn', 79; 'Prairie Dusk', 79; 'Prairie Fire', 79; 'Prairie Splendor', 78
HYBRIDS: European Hybrids, 69; 'Flathead Lake Hybrid', **70**, 71, 78–79; *Penstemon grandiflorus* × *P. nurranyanus*, 78, 79; *P. hartwegii* × *P. gentianoides*, 69; *P. hybridum*, 67, 69; *P.* × *johnsoniae*, 71
SPECIES: *Penstemon barbatus*, 67, 71; *P. campanulatus*, 67; *P. cardinalis*, 79; *P. cobaea*, 69; *P. gentianoides* var. *splendens*, 69; *P. hartwegii*, 69; *P. strictus*, 79
peonies, 2, 75, 203
Perry, Amos, 195, 197–98
Pertuzès, Dominique, 39, 45, 47
Pertuzès, François, 39, 45
Pethers, Thomas, 37, 53
petunias, 55, 137
Pfitzer, Wilhelm, 3, 75, **76**, 109
Pfitzer, Wilhelm, II, 77, 109
photoperiodism, 13

pinks. *See* carnations and pinks
Pins, Marquis de, 50
Pitcher & Manda Nurseries, 58–59
Planchon, J. E., 192
Plant Patent Act, 201
Plant Variety Protection Act, 11
Poeppig, Edward Friedrich, 193
Poinsett, Joel Roberts, **8**, 8–12
poinsettias, 5–17, 22; cultivation of, 13–14, 16; dissemination of, 5–6, 8–12, 16–17; grafting and, 6, 17; patent issues on, 11–12
    CULTIVARS: 'Annette Hegga Red', 17; 'Early Red', 14; 'Eckespoint Chianti', 16; 'Eckespoint Freedom', 14, 16; 'Eckespoint Lilo', 14, 16; 'Eckespoint Plum Pudding', 16; 'Eckespoint Prestige Red', 16; 'Eckespoint Shimmer Pink', 16; 'Eckespoint Winter Rose Dark Red', 14, 16; 'Henriette Ecke', 14, 15; 'Hollywood', 14; 'Indianapolis Red', 15; 'Mrs. Paul Ecke', 14, 15; 'Oak Leaf', 14; 'Paul Mikkelsen', 14, 16; 'St. Louis', 14; 'True Red', 14; 'V-14 Glory Angelikas', 17
    SPECIES: Peach poinsettia, **15**; *Poinsettia pulcherrima* (*Euphorbia*), 7, 8–9
Prescott, William Hickling, 9–10
Prestgard, Kristian, 104–6 (**105**), 111
Prince, William, 95
Pring, George H., **190**, 200–201

Ragonot-Godefroy, M., 177
Ramatuelle, Thomas d'Audibert de, 19, 21
Rand, Edward S., Jr., 96, 199
Randig, Martin R., 201–2
Redondo Carnation Company, 139
Reinelt, Frank, 203–4
Reydellet, Alexandre de, 39, 48
Reydellet, Auguste de, 38
Richardson, William, 170, 180
Rippy, J. Fred, 9
Robinson, William, 31, 32–33, 44, 60, 149, **151**, 189
Rodie, Hugh, 194
Roggli family, 176
roguing, 168–69
Ronaldson, James, 10–11
Root, Dave, 92
roses: *Rosa* 'Louise Odier', 182; 'Shell Pink Shawyer', 61; yellow, 61
Royal Botanic Gardens: Edinburgh, 179; Kew, 66, 192, 195, 200, 205
Royal Caledonian Horticultural Society, 180
Royal Horticultural Society, 26, 32, 69, 109, 125–27, 178, 195, 198
Royal Nursery, 122, 128–29, 152
Rozain-Boucharlat, Joseph, 39, 48
Russell, George, 100
Ryder, Samuel, 99

Sabine, Joseph, 24–25
Salisbury, Richard Anthony, 83

Salter, Alfred, 33
Salter, James, 57
Salter, John, 24, 33, **34**, 37, 40, 172, 184
Santel, M., 39
Say, Thomas, 10
Scharf, Alan, 79
Schmitt, M., 117, 119, 131
Schneevoogt (Dutch firm), 111
Schwanecke (German firm), 187
Seeba, Lina, 79
Seibert, Isabella, 190
Sharanova, Maria Fodorovna, 163–64
Sharma, Jagan ("Jaggi"), 176
Shelmire, Warren R., 139
Shepherd, Theodosia, 55, 99, 137
Shinners, Lloyd, 67
Siebold, Phillip von, 144
Simmons, W., 121
Skidelsky, Simon S., 57, 121, 132–34
Slocum, Perry D., 191, 202, 205
Small, Earl J., 14
Smith, Charles, 37
Smith, Elmer, 51, 59
Smith, Martin, 126
Smith, Nathan, 38
Société Française des Chrysanthémistes, 48
Société National de Horticulture Français, 47
Society of American Florists, 61
Souchet, Eugène, 83, 91, 95, 97
Spaargaren, J. J., 19
Spaulding, Thomas H., 59–60
Standish, John, 151–52
Stapeley Water Gardens, 194
Starr, Charles, 120, 121, 140
Stevens, George, 35
Stewart, Eugene Elmer, 107–8
Stewart, Robert N., 14
Stille, Charles, 9
St. John's Nurseries, 35
Stoke Newington Chrysanthemum Society, 24
Strawn, Robert Kirk, 203
Strong, William C., 97
Stuart, Charles, 179, 180
Sturtevant, Edmund T., 199
Sunningdale Nursery, 151–52
Suttons Seeds, 175

Teesdale, Charles Lennox Moore, 35
Theophrastus, 113
Thomas, Charles B., 202
Thomas, George Leicester, and family, 200, 204
Thompson, J. D., 132
Thompson, William, 69
Thomson, William, 169–70, 174, 177, 180
Thorburn, George, 73, 137
Thorpe, John, 60–61, 140
Torrey, John, 67
Totty, Charles H., 51, 61
Townsend, Stephen F., 83

Townsend-Purnell Act, 11
Tricker, William, 205
Trimardeau, Alexandre, 174, 176, 183, 186
Turner, Charles, 122–23, 128–29
Turner, William, 115

Uber, Ted, 201–2
Unwin, Frank, 92–93

Van Fleet, Walter, 89
Van Houtte, Victor, 2, 84, 85, 98, 108, 176, 192
Van Ness Nurseries, 201–2
Vaughans Nursery, 174
Vavilov, Nikolai, 163
Vawter, E. J., 122
Veitch, Harry, 149
Veitch, John, and family, 32, 35–36, 125, 152
Velthuys, Klaus, 93
Versailles Nursery, 33, 172
Vick, James, 97
Victoria and Albert, 83, 128
Viehmeyer, Glenn, 77, 77–79
Vilmorin-Andrieux & Cie., 48–49, 95, 183, 186
Viviand-Morel, Joseph Victor, 157
Volosenko-Valenis, Alexander Nicolaevich, 163, 164

Walcott, Henry Pickering, 61–62
Wallace, Henry, 106
Wallich, Nathaniel, 151
Ward, Charles Willis, 117, 120, 134, 141
Warren Nursery, 32
Wasuwat, Slearmlarp, 199
Waterer, Henry, 62–63
water lilies and lotuses, 189–205; classification of, 191–94, 203; dissemination of, 189, 191; Monet and, 198
  CULTIVARS: 'Afterglow', 202; 'Angel Wings', 202; 'Aviator Pring', 201; 'Blue Beauty', 205; 'Blue Lotus of the Nile', **193**; 'Blue Lotus of India', 193; 'Charlene Strawn', 203; 'Charles Thomas', 202; 'Conqueror', 198; 'Crystal Lingg'; 200; 'Dorothy Lamour', 204; 'Ellisiana', 198; 'Evelyn Randig', 201; 'Froebeli', 199; 'Fulva', 198; 'George Huster', 194; 'Gladstone', 198; 'Green Smoke', 202; 'Gulyanee' 199; 'Larp Prasert', 199; 'Maggie Bell Slocum', 202; 'Margaret Mary', 204; 'Mrs. C. W. Thomas', 204; 'Nangkwang Chapoo 1', 199; 'O'Marana', 194; 'Pearl of the Pool', 202; 'Pennsylvania', 205; 'Primlarp', 199; 'St Louis', 201; 'Texan Shell', 200
  HYBRIDS: *Nymphaea kewensis*, 192; *N. 'Ortgesiana-rubra'*, 192
  SPECIES: *Nelumbo lutea*, 192; *N. nucifera*, 192–93; *Nymphaea alba*, 191; *N. alba* var. *rubra*, **197**, 198–99; *N. bisseti*, 194; *N. capensis*, 191; *N. coerulea*, 193; *N. colorata*, 191; *N. dentata*, 194; *N. devoniensis*, 192; *N. gigantea*, 191; *N. lotus*, 191–92; *N. mexicana*, 197; *N. nouchali*, 193; *N. odorata*, 191, 197; *N. ortgesiana*, 192; *N. sturtevanti*, 194; *N. rubra*, 191–92; *N. tetragona*, 191, 195; *Victoria amazonica*, 189, 192, 193–94, 195, 199, 205; *V. cruziana*, 193–94, 205
Way, David, 69
Weber, Henry, and sons, 140–41
Weeks, John, 30
weigelas, 2
Wells, William, 36
Wharton, Edith, 61
Wheeler, Isaac, 24
Whitbourn, Francis, 125, 126
Willdenow, Karl, 9
William and Mary, 23
Willson, Eleanor, 151–52
Witterstaetter, Richard, 57, 121, **141**
Wittrock, Veit Brecher, 167–68, 172, 177, 183, 187
Woolfenden, Ken, 158
Wrede, H., 187
Wright, Joan, 111

Yoder Brothers, 14
Young, Brian, 37

Zeiger Brothers, 17
Ziemann, Johan Heinrich, 185